T0201082

Connections

THE WILEY BICENTENNIAL–KNOWLEDGE FOR GENERATIONS

Each generation has its unique needs and aspirations. When Charles Wiley first opened his small printing shop in lower Manhattan in 1807, it was a generation of boundless potential searching for an identity. And we were there, helping to define a new American literary tradition. Over half a century later, in the midst of the Second Industrial Revolution, it was a generation focused on building the future. Once again, we were there, supplying the critical scientific, technical, and engineering knowledge that helped frame the world. Throughout the 20th Century, and into the new millennium, nations began to reach out beyond their own borders and a new international community was born. Wiley was there, expanding its operations around the world to enable a global exchange of ideas, opinions, and know-how.

For 200 years, Wiley has been an integral part of each generation's journey, enabling the flow of information and understanding necessary to meet their needs and fulfill their aspirations. Today, bold new technologies are changing the way we live and learn. Wiley will be there, providing you the must-have knowledge you need to imagine new worlds, new possibilities, and new opportunities.

Generations come and go, but you can always count on Wiley to provide you the knowledge you need, when and where you need it!

WILLIAM J. PESCE
PRESIDENT AND CHIEF EXECUTIVE OFFICER

PETER BOOTH WILEY
CHAIRMAN OF THE BOARD

Connections

Patterns of Discovery

H. Peter Alesso and Craig F. Smith

WILEY-INTERSCIENCE
A John Wiley & Sons, Inc., Publication

Copyright © 2008 by John Wiley & Sons, Inc. All rights reserved.

Published by John Wiley & Sons, Inc., Hoboken, New Jersey
Published simultaneously in Canada

No part of this publication may be reproduced, stored in a retrieval system, or transmitted in any form or by any means, electronic, mechanical, photocopying, recording, scanning, or otherwise, except as permitted under Section 107 or 108 of the 1976 United States Copyright Act, without either the prior written permission of the Publisher, or authorization through payment of the appropriate per-copy fee to the Copyright Clearance Center, Inc., 222 Rosewood Drive, Danvers, MA 01923, 978-750-8400, fax 978-750-4700, or on the web at www.copyright.com. Requests to the Publisher for permission should be addressed to the Permissions Department, John Wiley & Sons, Inc., 111 River Street, Hoboken, NJ 07030, 201-748-6011, fax 201-748-6008, or online at http://www.wiley.com/go/permission.

Limit of Liability/Disclaimer of Warranty: While the publisher and author have used their best efforts in preparing this book, they make no representations or warranties with respect to the accuracy or completeness of the contents of this book and specifically disclaim any implied warranties of merchantability or fitness for a particular purpose. No warranty may be created or extended by sales representatives or written sales materials. The advice and strategies contained herein may not be suitable for your situation. You should consult with a professional where appropriate. Neither the publisher nor author shall be liable for any loss of profit or any other commercial damages, including but not limited to special, incidental, consequential, or other damages.

For general information on our other products and services or for technical support, please contact our Customer Care Department within the United States at 877-762-2974, outside the United States at 317-572-3993 or fax 317-572-4002.

Wiley also publishes its books in a variety of electronic formats. Some content that appears in print may not be available in electronic formats. For more information about Wiley products, visit our web site at www. wiley.com.

Library of Congress Cataloging-in-Publication Data:

Alesso, H. P.
Connections: Patterns of discovery / by H. Peter Alesso and Craig F. Smith ; with foreword by James Burke.
p. cm.
Includes bibliographical references and index.
ISBN 978-0-470-11881-8 (cloth)
1. Information technology. 2. Discoveries in science. I. Smith, C. F. (Craig Forsythe), 1950– II. Title.
T58.5.A54 2008
004—dc22

2007017344

Printed in the United States of America

10 9 8 7 6 5 4 3 2 1

Contents

Foreword

Every since the flint axe triggered the Paleolithic Revolution over two million years ago we have lived by a culture of scarcity. Technology shortfall placed innovative capability in the hands of only a few: those who provided their hunter-gatherer communities with tools for hunting and butchering; then those who could read and write; then, much later, the small number adept in the mysteries of science. For most of history the vast majority of the community, illiterate and uneducated, were excluded.

As time passed, institutions emerged in response to the requirements generated by innovation. The organizational demands of the Paleolithic hunt shaped our top–down command structures. The Agricultural Revolution of ten thousand years ago provisioned the first cities and triggered the emergence of our legal and political systems. The printing press established our national languages and created the modern state. The 19th century Industrial Revolution introduced our first grade schools to train farmhands for the factory.

Every social institution in the modern world was set up in the past, with the limited tools of the past, in order to solve the problems of the past. Few of these institutions will survive what lies ahead, as the new Information Revolution gathers speed and strength, and drives accelerating rates of innovation faster than the old ways can manage. If we are to prepare for the coming radical social changes, at every level from the personal to the global, we must find ways better to predict, and to organize ourselves accordingly.

To search for indicators in this endeavor, the authors of this book look first to the past (since there is nowhere else to look and, as every traveler knows, you only know where you're going if you know where you've been). In their fascinating analysis of the recent history of information technology, Peter

Alesso and Craig Smith reveal the patterns in discovery and innovation that have brought us to the present tipping point.

They penetrate the esoteric language of science to illustrate entertainingly, comprehensively, and, above all, clearly how science and technology even of the most arcane and complex kind involve stuff we can all understand. Their cogent explanation of the research behind fundamental advances in information technology also features mistakes, visionaries, class war, genius, garages, refugees, dropouts, serendipity, method, and just plain hard work.

The book's conclusion looks ahead to the end of the culture of scarcity. A generation from now every individual will have personally tailored access to the whole of knowledge. And thanks to ubiquitous, invisible, networked devices, each of us will also live within an intelligent personal space that will interact intelligently with an equally intelligent environment. We will be informed and enfranchised as never before. No present-day social organization was built to deal with this.

We now face social change as profound as that which confronted the users of the first Paleolithic flint axe. The sooner we *all* begin to think about how we got here, and where we're going, the better. This exciting book is an essential first step.

James Burke

James Burke has an Oxford M.A. and holds three honorary doctorates for his work in communication. He has written eleven books (including *Connections*, *The Day the Universe Changed*, and most recently *American Connections*) and has produced many prize-winning television documentary series, including the landmark ten-part BBC/PBS "Connections," in which Burke as presenter followed a series of seemingly-unrelated people and events to illustrate the serendipitous nature of technological change. His latest project, due online in 2008, is a knowledge-web of 2,500 historical personalities linked over 30,000 ways. On the U.S. lecture circuit, Burke is a keynote speaker for organizations such as the Smithsonian, Microsoft, NASA, MIT, the European Parliament, and many universities and colleges. The *Washington Post* called him "one of the most intriguing minds in the Western world."

Acknowledgments

It is a pleasure to acknowledge the inspiration for this book—James Burke and his *Connection* series. We found Burke's work to be a brilliant examination of ideas, inventions, and events affecting people and society. We enjoyed every facet of his peerless presentation.

We would like to acknowledge the excellent reference material of David A. Vise and Mark Makseed's book *The Google Story*.

We also recognize the outstanding predictions of the researchers at Elon University and the Pew Internet Project (`http://www.elon.edu/e-web/predictions/whystudyfuture.xhtml`), who conducted a survey of leaders, asking them to share their expectations for the future.

In addition, we concede that the biographies and personal stories of many of the inventors included in this book have been told and retold many times. While we have endeavored to present this material as uniquely and freshly as possible, some resemblance to previous work cannot be avoided simply because the basic facts—birth dates, places of work, parents, education, and major accomplishments—remain the same.

We also appreciate the style ideas and suggestions of Chris Alesso and the review efforts and suggestions of Kathleen Smith.

Finally, we thank George J. Telecki, associate publisher at Wiley-Interscience, and his staff, including Rachel Witmer, Melissa Valentine, and Rosalyn Fankas for their tireless efforts during the preparation of this work.

H. Peter Alesso
Craig F. Smith

Organization of this Book

This book is about connections. Each chapter examines the respective connections between information, circuits, chips, processes, computers, networks, devices, ubiquitous computing, the ubiquitous Web, and ubiquitous intelligence. And each chapter highlights unique lessons in the patterns of discovery.

Chapter 1 presents the hero's journey of two young inventors using inspiration and perspiration to create the world's best search engine, Google. In Chapter 2, the growth of the Information Superhighway is linked to the succeeding generations of proof of principle inventions fueling Moore's Law: the vacuum tube, the transistor, and the microprocessor.

In Chapter 3, the story of the personal computer reveals how the incubator research center of Xerox PARC led the breakthrough in six technologies only to fail commercially in contrast to the success of Apple and IBM PC following the perspiration/inspiration pattern. Chapter 4 shows the artistry of software development giving rise to five generations of technology, but failing to duplicate hardware's success under Moore's Law. In addition, it illustrates top–down (command) versus bottom–up (emergent) behavior in relation to proprietary and open software systems. Then in Chapter 5, the Ethernet story describes connecting networks around the world.

Chapter 6 tells the Internet story with a discussion of Tim Berners-Lee's contribution to the development of the World Wide Web. It highlights open simple standards as a powerful force for global behavior. Chapters 7, 8, and 9 extend the discussion of connections by considering ubiquitous computing, the ubiquitous Web, and ubiquitous intelligence. Finally, in Chapter 10, we look to futurist Ray Kurzweil for inspiration on the future of discovery patterns as the rate of technological change accelerates.

Introduction

The most remarkable discovery made by scientists is science itself.
—Jacob Bronowski[1]

Our past is a tapestry—rich with dramatic experiences of discovery. The patterns that emerge from its fabric typecast inventions both individually and collectively. Extending the tapestry's quilted pattern, we can forecast coming innovations.

As the past flows through the present in this way, it unfolds the remarkable stories of inventors—revealing the "big picture" of discovery and foreshadowing the next generation of information technology.

STORIES

Stories offer the drama and excitement of human experiences. Our thoughts and memories are organized around stories. They offer imaginative narratives to stimulate ideas and help us understand the world.

In 1949, Joseph Campbell wrote *The Hero with a Thousand Faces*, in which he argued that myths—across all time and all cultures—contain the same basic elements or "archetypes."

Campbell claimed that all stories are expressions of the same story pattern, which he called the "hero's journey." The hero's journey has three parts: departure, initiation, and return. It begins as the hero hears a pleading call. At first, he refuses the call, but eventually he responds to it by departing and crossing

[1]J. Bronowski, *The Creative Process*, Scientific American, W. H. Freeman and Company, New York, 1982.

Connections: Patterns of Discovery By H. Peter Alesso and Craig F. Smith
Copyright © 2008 John Wiley & Sons, Inc.

a threshold into a new realm. As he initiates his journey, he faces great challenges and he matures in understanding with the help of mentors he encounters along the way. Finally, he returns as the master, committed to improve the world. This structure underlies stories from Homer's *Odyssey* to George Lucas's *Star Wars*.

The hero's journey has had incredible ramifications for storytellers in general. But, as central as stories are to the theater of human behavior, tales of scientific discovery are frequently portrayed as anecdotal and unique, rather than illuminating and connected.

Twenty-three centuries ago, the famous Greek mathematician and scientist Archimedes undertook the task of determining if his king's crown was made of pure gold or if it had been fraudulently alloyed with less expensive metals. He knew that if baser metals had been used, the crown would be less dense than if it were of pure gold. While bathing in the public baths, it dawned on him that he could measure the volume of the solid crown by its displacement of water. From the volume and the weight of the crown, he could determine its density. It is said that following this moment of insight, he ran naked from the baths and into the streets shouting "Eureka!" As a result of this tale, scientists are often viewed as dispassionate logical technicians who slavishly repeat experiments until something unexpected happens and then they exclaim Eureka!

Actually, there is a great deal more to the story of scientific discovery, but "seeing the big picture" is not easy. It requires a grasp of the relationship between relationships. And forecasting requires conceptual tools such as the S-shaped curve, the Delphi method, and archetypal patterns.

The truth is that the biggest successes come when a scientist takes a leap of imagination out of *what is probable* into *what might be possible*. As scientists seek such understanding in order to control their surroundings, understanding comes from three remarkable human characteristics: discovery, invention, and creativity.

We associate discovery with finding something that is already there—like Columbus discovering America. Invention is considered the product of ingenious thought, such as Alexander Graham Bell's invention of the telephone. In contrast, creativity is more the result of a single mind producing a unique piece of art, such as a play by Shakespeare.

There are discoveries in science that are similar to Columbus's discovery, where something is found that was already there—an example is the principle of buoyancy discovered by Archimedes. And there are ingenious inventions, like Bell's invention of the telephone, which combine a set of known principles—like sound and electricity. But science makes room for creativity as well. A fact may be discovered, but a theory is invented by using a creative mind. Bell once said that "the most successful men in the end are those whose success is the result of steady accretion."[2]

[2]http://www.leadershipvictoria.org/resources/quotes/quotes_otherB.htm.

Science and the arts have flourished together by relying on creativity. Their search is for the unity and simplicity in nature. Thus the world view of the artist, as well as scientists, is forward looking—each forecasting the progress of man.

In *Connections: Patterns of Discovery*, we identify and analyze three archetypal patterns of discovery. The first and rarest is the *Serendipity Pattern*, where pure chance plays a role in a discovery. In the second, the *Proof of Principle Pattern*, a scientist uses a known scientific phenomenon to invent a proof of principle application. The third and most common, the *1% Inspiration and 99% Perspiration Pattern*, reflects Edison's method, where a scientist starts out with a known phenomenon and an existing proof of principle experiment and by innovative ideas and hard work goes on to invent a new commercially competitive product.

We will find in Chapter 1 that Larry Page and Sergey Brin's invention of the Google ranking algorithm that could be efficiently deployed using a specialized software and hardware environment illustrates the *1% Inspiration and 99% Perspiration Pattern*.

But patterns of discovery can also be explored collectively, leading to the creation of an entire industry. For example, discoveries that influence succeeding generations of technology can follow a "pattern of patterns."

The Moore's Law story in Chapter 2 illustrates just such a case. It is a trilogy of stories—the vacuum tube, the transistor, and the microprocessor—each contributing to the development of the computer. This collection of stories for three inventions reveals a collective pattern.

Another aspect of patterns of discovery can be observed when competing discoveries vie for dominance. An example of this is illustrated in Chapter 3 by the development of the personal computer. The Xerox Palo Alto Research Center (PARC) introduction of the early concept for a computer workstation, known as Alto, followed a series of *Proof of Principle Patterns.* But it was a commercial failure. The success of the competitors, the Apple II and the IBM PC, followed inspiration and perspiration.

For us, the past is prologue; our stories unfold from historic perspective but soon reveal patterns that foreshadow the next generation of technology centered on new capabilities in areas such as artificial intelligence, advanced networking, and information systems. They challenge the reader to think of the consequences of extrapolating trends, such as Moore's Law, to achieve machine intelligence, or retrench in the face of physical limitations. The discourse leads to questions such as: What is the software equivalent of Moore's Law?

From this perspective, the book draws the big picture for the Information Revolution's innovations in chips, computers, devices, software, and networks. The goal is to forecast the elements of ubiquitous intelligence (UI), where everyone is connected to devices with access to artificial intelligence (AI).

The three step process of creating ubiquitous intelligence starts with the process of ubiquitous computing populating the world with microchip devices everywhere. Then the ubiquitous Web will connect and control these devices on a global scale. The final step comes when artificial intelligence reaches the capability of self-managing and regulating devices seamlessly and invisibly within the environment—achieving ubiquitous intelligence.

CONNECTING INFORMATION

Beginning in the mid-20th century, there was an explosion in technology innovation, which characterized the start of the Information Revolution. For the first time, ordinary people had real power over information production and dissemination. Computer technology offered reduced information costs, allowing individuals to buy, sell, share, and even create their own information.

As a result, the World Wide Web became the "grim reaper" of information inefficiency. And the Web will clearly become more powerful as it acquires greater intelligent capabilities.

So where better to start in our search for patterns of discovery, than with the world's greatest search engine, Google. The story of Google is the story of two inventors and the most remarkable Internet success of our time.

Today, searching the Web is an essential capability whether you are sitting at your desktop PC or wandering the corporate halls with your wireless PDA. Consequently, many companies have entered the business of providing commercial search engines, and the practice of Web search is now an essential capability demanded by Web users. Clearly, Google has become one of the most recognizable names in the modern corporate world.

Google was founded by two graduate students at Stanford University—Larry Page and Sergey Brin. Larry and Sergey were two young men, kids really—brilliant, well-educated, excitable, enthusiastic, argumentative. Larry was fond of saying that he had a "healthy disregard of the impossible." It was 1995 when they met. They argued incessantly. You would think they wouldn't get along, but instead they bonded and formed a lasting friendship. And they had a mutual goal—to download the World Wide Web and search it at their leisure. The effort turned out to be more formidable then they imagined. But the result was Google.

In Chapter 1, we follow the journey of Page and Brin as they find inspiration, invent their Googleware technology, and dream of what Larry Page calls "perfect search."

In subsequent chapters, we will continue to follow the journeys of key scientists and technologists as they invent innovative devices and features that form the foundation of the Information Revolution. The innovations include new chip technologies that power ever more capable computers, connecting expanding networks, reaching smaller ubiquitous devices, and running artificially intelligent applications. The goal of the Information Revolution will be ubiquitous intelligence (UI) that will endeavor to achieve Larry Page's dream of perfect search.

PATTERNS

What does it take to recognize patterns of discovery in the stories of scientists? To some extent, becoming a master of discovery patterns is like becoming a master chess player.

The process of becoming a chess master consists of several steps. First, learn the rules including the names of pieces, legal movements, chess board geometry, and orientation. Second, learn the basic principles including the relative value of the pieces, the strategic value of the center squares, and the power of a threat. Third, study the games of the masters, including those games containing patterns that must be understood and applied.

Similarly, recognizing patterns of discovery requires the following analogous steps:

1. *Learn the rules*—about talent, knowledge, and resources skillfully applied.
2. *Learn the principles*—serendipity, proof of principle, and inspired exertion.
3. *Study the designs of masters*—find the patterns of master inventors such as Thomas Edison.

By taking this approach, it is clear how patterns play a vital role in developing scientific patterns of discovery. But understanding recognized patterns is just the beginning of the process of thinking in terms of using patterns to create innovation.

PATTERNS OF DISCOVERY

There are creators who astonish us by the sheer volume of their prolific contributions. Thomas Edison is such a man. He amassed 1093 patents—an unrivaled achievement. He began his career in telegraphy and continued to produce communication inventions in a seemingly endless parade of contributions to telephone, radio, phonograph, photography, and motion pictures. But he was a genius who could not be contained. He went on to develop the electric light, dynamo, motor, electric distribution system, and more.

His efforts were not only prodigious they were catholic in their methodology. So pervasive are the creative contributions of Edison, it is fair to say that if there are patterns of discovery, then Edison made the molds.

We propose that the hero's journey of our book be centered on three archetypes or patterns of discovery, *Serendipity, Proof of Principle*, and the *1% Inspiration and 99% Perspiration*.

The Serendipity Pattern is the first and rarest of the patterns, where pure chance plays a role to produce a discovery even while the inventor is engaged in a prolonged investigation. Serendipity is the detection of things through the faculty of making fortunate and unexpected discoveries by accident. Many diverse discoveries such as penicillin, X-rays, and Teflon were the result of accident and sagacity.

Pasteur made great breakthroughs in chemistry, microbiology, and medicine; he clearly recognized the role that serendipity plays in discovery when he said, "In the fields of observation, chance favors only the prepared mind."[3]

Nobel Prize winner Paul Flory said, "Significant inventions are not mere accidents. The erroneous view [that they are] is widely held, and it is one that the scientific and technical community, unfortunately, has done little to dispel. Happenstance usually plays a part, to be sure, but there is much more to invention than the popular notion of a bolt from the blue. Knowledge in depth and in breadth are virtual prerequisites. Unless the mind is thoroughly charged beforehand, the proverbial spark of genius, if it should manifest itself, probably will find nothing to ignite."[4]

People should be encouraged to be flexible in their thinking and interpretations. Without such flexibility, it can be expected that people will judge new information in terms of their preconceived notions and potentially miss the observance of new trends or insights that would constitute discovery if only they were recognized as such.

Serendipity can benefit inventors if they are well prepared. The American physicist Joseph Henry paraphrased Pasteur when he said, "The seeds of great discovery are constantly floating around us, but they only take root in minds well prepared to receive them."[5]

Any of the accidental discoveries leading to invention by serendipity could have gone unnoticed. Instead, because of the wisdom of the individuals who encountered the accident, we have an explanation of X-rays for medical diagnosis; the "miracle drug" penicillin and its descendants; and Teflon for heart valves. We can be sure that accidents will continue to happen and with a prepared mind we can expect them to lead to discoveries.

Edison's discovery of the phonograph is an example of the serendipity pattern. When Edison was experimenting with the telegraph transmitter in an effort to improve its design, he noticed a sound emanating from the equipment that had a certain resemblance to human speech. Although this was a surprise, Edison's prepared mind immediately grasped its significance. He determined to pursue his observation with the intent of finding out whether or not meaningful sounds in the form of human speech could be recorded. He began his investigation by attaching a needle to the diaphragm of a telephone receiver, and he subsequently refined his experiment using a stylus on a tinfoil cylinder. This succeeded when he was able to record and play back the message, "Mary had a little lamb."

[3]L. Pasteur, "1854 Lecture at the University of Lille," in *A Treasury of World's Great Speeches*, H. Peterson (ed.), Spencer Press, Chicago, 1954.
[4]P. Florey, on receipt of the ACS Priestley Medal, quoted in *Chemistry in Action, Chemistry and Chance: Part 1*, by P. E. Childs, `http://www.ul.ie/~childsp/CinA/issue50/chance.html`.
[5]"Anatomy of Discovery," September 29, 2006. `http://www.physorg.com/news78754375.html`.

Here a serendipitous accident took place before a genius engaged in a serious, long-term investigation that prepared him to recognize the implications of the "accident." On the other hand, the public's perception of the Serendipity Pattern could best be captured by a 1970 Hershey Food Corp. commercial where a person walks along absent-mindedly munching a chocolate bar, while another person dreamily strolls by eating peanut butter. The two collide.

"Hey, you got peanut butter on my chocolate," says one.

"And you got chocolate on my peanut butter," says the other.[6]

They sample the result and, Eureka! they discover Reese's Peanut Butter Cups.

Sometimes the most useful ideas come from simple combinations of existing ideas. Nevertheless, this type of invention is rare and by its unexpected nature defies forecasting.

S. Harris's cartoon (reprinted with permission of ScienceCartoonPlus.com) encapsulates the difficulty of planning for serendipity.

"I THINK YOU SHOULD BE MORE EXPLICIT HERE IN STEP TWO."

[6] D. H. Pink, *A Whole New Mind*, Riverhead Books, New York, 2005.

In the second type of pattern of discovery, a scientist uses a known phenomenon to invent a proof of principle application. In this case a known phenomenon stimulates exploration for a particular application. If the inventor is able to demonstrate the phenomenon accomplishing a particular application, we consider that he/she has completed a Proof of Principle Pattern.

The second type of pattern of discovery is illustrated by John Fleming's vacuum tube diode.

Before Fleming came on the scene, Thomas Edison had been experimenting with light bulbs and in 1883 he found that he could detect electrons flowing through the vacuum to a metal plate mounted inside the bulb. This was a serendipitous discovery by Edison, who did not set out to invent this process. The phenomenon subsequently became known as the "Edison Effect."

Later, in 1904, while Fleming was investigating the Edison Effect, he discovered the effect could be used not only to measure flowing electrons but also to detect radio waves and to convert them to electricity.

Fleming introduced his device, known as the diode, that he constructed by simply adding an additional electrode inside an incandescent light bulb. The Fleming diode was capable of converting alternating current into direct current and found great application in power supplies for electronic equipment; however, in a more immediate and important application, it could be used as a detector for the very weak radio signals that were characteristic of the newly introduced wireless telegraph.

Fleming's work is an example of the second type of pattern of discovery. Here Fleming used a known phenomenon—the Edison Effect—to invent a proof of principle application—the vacuum tube diode.

And the third and most common pattern of discovery is Edison's 1% Inspiration and 99% Perspiration Pattern, where a scientist starts out with a known phenomenon and an existing proof of principle for a particular application and goes on to invent a commercially competitive product. This can be illustrated by Edison's quest for the incandescent light.

In 1878, Edison became intrigued with electric lighting. Until then only arc lighting with Jablochkoff's "candles" for spotlighting in theaters had been achieved. In 1808, Humphrey Davy produced an arc as a flash of light. In 1845, J. W. Starr heated a rod of carbon in a vacuum, and in 1860, J. Swan experimented with a strip of paper, but because his vacuum was inadequate his results were not useful.

Edison arrived at the right time. The phenomenon of producing light from heat had been established and some proof of principle experiments had established electric lighting as a potential application. Edison applied his inspiration and perspiration approach. It was his inspired insight that enabled him to see that a high resistance filament, in a parallel circuit using newly developed high vacuum technology, could produce a commercial incandescent light.

The invention and development of the electric lighting system followed a pattern of rapid growth (an S-shaped curve) as it replaced the gas lighting system and expanded the overall market.

This was the pattern of discovery that rewarded Edison's years of perspiration with the incandescent light bulb. We shall see that this is an often repeated pattern behind many scientific endeavors.

But patterns of discovery should not be explored only as isolated events. They are also informative when examined in combinations that influence succeeding generations of technology.

In this book, our stories enrich the tapestry of discovery. They reveal patterns in the various fields of information technology and they help forecast trends in innovations.

FORECASTING TOOLS

The forecasting of future relationships requires a systematic look into the prospects of science, technology, and the economy to identify areas of strategic opportunity.

But forecasting can be a dangerous enterprise; it is far easier to get it wrong than to get it right. Lessons can be taken from past efforts to forecast technology. In a recent interesting review[7] of a forty-year-old forecasting study,[8] Richard E. Albright commented on the 100 technical innovations identified as being considered very likely to be developed in the last third of the 20 century. While fewer than 50% of the predicted innovations were considered "good and timely," Albright found that the accuracy rate for the areas of computers and communications rose to about 80%.

Furthermore, for current predictions, Albright concluded that "we should look for sustained and continuing trends in underlying technologies, where increasing capabilities enable more complex applications and declining costs drive a positive innovation loop, lowering the cost of innovation and enabling wider learning and contributions from more people, thus sustaining the technology trends."

The primary tools of forecasting include the S-shaped curve, envelope curves, trend extrapolation, the Delphi method, and archetypal patterns.

The growth of a new technological capability typically follows an S-shaped curve, which can be divided into three stages. The first stage is slow initial growth, where the new technology has to prove its superiority over existing competitors. Once this is demonstrated, a period of rapid growth follows. Finally, its growth is limited by technological or socioeconomic competition and levels off asymptotically toward an upper limit.

The S-shaped curve is a pattern of development that illustrates progression of many inventions. For example, Thomas Edison developed electrical

[7]R. E. Albright, "What Can Past Technology Forecasts Tell Us About the Future?," *Technological Forecasting and Social Change* 69(5): 443–464 2002.

[8]H. Kahn and A. Wiener, *The Year 2000, A Framework for Speculation on the Next Thirty-Three Years*, MacMillan Publishing Company, London, 1967.

appliances for the home and the factory. Many of these devices were based on analog signal processing and they opened the door to independently powered machines. But this simple paradigm shift took nearly fifty years to come to practical fruition as the adoption and utilization of independently powered analog machines followed an S-shaped curve through their introduction, rapid growth, and saturation stages.

The Delphi method develops forecasts based on the opinion of a panel of independent experts through repeated rounds of review.

In addition, patterns of discovery revealed in the scientist's journey offer insights that emerge to create a tapestry for future innovations. We recognize archetypal patterns such as "patterns of discovery" and "patterns of patterns" to support our forecasts of technology change.

THE BIG PICTURE

Seeing the big picture is the killer-app in thinking and analysis. It requires the ability to grasp the relationship between relationships. It requires the mind-set of a pattern recognizer. Take special note of the space between the "E" and the "x" in the logo below. If you look carefully, you will recognize that the space between these letters forms an arrow.

That's a powerful pattern called *negative space*. Negative space is often the part of the big picture that we overlook.

There are several technology extrapolation models that can be used for predictions of varying quality. The application of Moore's Law in International Technology Roadmap for Semiconductors (ITRS)[9] is a rather successful one.

B. V. Gnedenko and A. N. Kolmogorov, in their 1954 work *Limit Distributions for the Sum of Independent Random Variables*,[10] noted that emergent order is constrained by randomness in nonlinear complex systems. Self-organization theory in complex systems research adds insight.

[9]"International Technology Roadmap for Semiconductors (ITRS)," http://www.itrs.net/.

[10]B. V. Gnedenko and A. N. Kolmogorov, *Limit Distributions for Sums of Independent Random Variables*, Addison-Wesley, Boston, 1954.

However, complex developing systems can become predictable in at least two ways. If the systems are cyclic in their behavior, the patterns of these cycles can be subject to discernment; and in the case of more complex systems, behavior can be recognized as emergent in nature. In this case as well, the patterns of the emergent behavior can be identified.

In the stories of inventors and inventions, we will find patterns of discovery. From this we will highlight the course of technology development as a forecast of potential innovation.

FORECASTS

After a review of historical developments and emerging strategic opportunities, the resultant patterns offer us insights into big picture relationships that help illuminate the next generation of technology in a variety of fields, such as microchips, devices, software, and networks. In subsequent chapters of this book, we will attempt to forecast the future by interpreting the patterns we find.

We present the patterns of discovery that produced Moore's Law and we explore the question: What is the software equivalent of Moore's Law?

The patterns challenge the reader to think of the consequences of extrapolating trends, such as how Moore's Law could reach machine intelligence, or retrench in the face of physical limitations.

From this perspective, the book draws the big picture for the Information Revolution's innovations in chips, devices, software, and networks. The goal is ubiquitous intelligence (UI), where everyone is connected to devices with access to artificial intelligence (AI)—offering what Google founder Larry Page calls "perfect search."

1

Connecting Information

The ultimate search engine would understand exactly what you mean and give back exactly what you want.
—Larry Page[1]

We live in the information age. As society has progressed into the postindustrial era, access to knowledge and information has become the cornerstone of modern living. With the advent of the World Wide Web, vast amounts of information have suddenly become available to people throughout the world. And searching the Web has become an essential capability whether you are sitting at your desktop PC or wandering the corporate halls with your wireless PDA. As a result, there is no better place to start our discussion of *connecting information* than with the world's greatest search engine—Google.

Google has become a global household name—millions use it daily in a hundred languages to conduct over half of all online searches. As a result, Google connects people to relevant information. By providing free access to information, Google offers a seductive gratification to whoever seeks it. To power its searches, Google uses patented, custom-designed programs and hundreds of thousands of computers to provide the greatest computing power of any enterprise.

Searching for information is now called *googling*, which men, women, and children can perform over computers and cell phones. And thanks to small targeted advertisements that searchers can click on for information, Google has become a financial success.

In this chapter, we follow the hero's journey of Google founders Larry Page and Sergey Brin as they invent their Googleware technology for efficient con-

[1]M. Prather, "Ga-Ga for Google," *Entrepreneur Magazine*, April 2002.

Connections: Patterns of Discovery By H. Peter Alesso and Craig F. Smith
Copyright © 2008 John Wiley & Sons, Inc.

nection to information, then go on to become masters in pursuit of their holy grail—perfect search.

THE GOOGLE STORY

Google was founded by two Ph.D. computer science students at Stanford University in California—Larry Page and Sergey Brin. When Page and Brin began their hero's journey, they didn't know exactly where they were headed.

It is widely known that, at first, Page and Brin didn't hit it off. When they met in 1995, 24-year-old Page was a new graduate of the University of Michigan visiting Stanford University to consider entering graduate school; Brin, at age 23, was a Stanford graduate student who was assigned to host Page's visit. At first, the two seemed to differ on just about every subject they discussed. They each had strong opinions and divergent viewpoints, and their relationship seemed destined to be contentious.

Larry Page was born in 1973 in Lansing, Michigan. Both of his parents were computer scientists. His father was a university professor and a leader in the field of artificial intelligence, while his mother was a teacher of computer programming. As a result of his upbringing in this talented and technology-oriented family, Page seemed destined for success in the computer industry in one way or another.

After graduating from high school, Page studied computer engineering at the University of Michigan, where he earned his Bachelor of Science degree. Following his undergraduate studies, he decided to pursue graduate work in computer engineering at Stanford University. He intended to build a career in academia or the computer science profession, building on a Ph.D. degree.

Meanwhile, Sergey Brin was also born in 1973, in Moscow, Russia, the son of a Russian mathematician and an economist. His entire family fled the Soviet Union in 1979 under the threat of growing anti-Semitism and began their new lives as immigrants in the United States.

Brin displayed a great interest in computers from an early age. As a youth, he was influenced by the rapid popularization of personal computers and was very much a child of the microprocessor age. He too was brought up to be familiar with mathematics and computer technology, and as a young child, in the first grade he turned in a computer printout for a school project. Later, at the age of nine, he was given a Commodore 64 computer as a birthday gift from his father.

Brin entered the University of Maryland at College Park, where he studied mathematics and computer science. He completed his studies at the University of Maryland in 1993, and was awarded a Bachelor of Science degree. Following his undergraduate studies, he was given a National Science Foundation fellowship to pursue graduate studies in computer science at Stanford University. Not only did he exhibit early talent and interest in mathematics and computer science, he also became acutely interested in data management and networking as the Internet was becoming an increasing force in American society. While at Stanford, he pursued research and prepared publications in the areas of data mining and pattern extraction. He also wrote software to convert scientific papers written in TeX, a cross-platform text processing language, into HyperText Markup Language (HTML), the multimedia language of the World Wide Web.

Brin successfully completed his master's degree at Stanford. Like Page, Brin's intent was to continue in his graduate studies to earn a Ph.D., which he also viewed as a great opportunity to establish an outstanding academic or professional career in computer science.

The hero's journey for Page and Brin began as they heard the call—to develop a unique approach for retrieving relevant information from the voluminous data on the World Wide Web.

Page remembered, "When we first met each other, we each thought the other was obnoxious. Then we hit it off and became really good friends. . . . I got this crazy idea that I was going to download the entire Web onto my computer. I told my advisor it would only take a week . . . So I started to download the Web, and Sergey started helping me because he was interested in data mining and making sense of the information."[2]

Although Page initially thought the downloading of the Web would be a short-term project, taking a week or so to accomplish, he quickly found that the scope of what he wanted to do was much greater than his original estimate. Once he started his downloading project, he enlisted Brin to join the effort. While working together the two became inspired and wrote the seminal paper entitled *The Anatomy of a Large-Scale Hypertextual Web Search Engine*.[3] It explained their efficient ranking algorithm, PageRank.

Brin said about the experience, "The research behind Google began in 1995. The first prototype was actually called BackRub. A couple of years later, we had a search engine that worked considerably better than the others available did at the time."[4]

This prototype listed the results of a Web search according to a quantitative measure of the popularity of the pages. By January 1996, the system was able to analyze the "back links" pointing to a given website and from this quantify the popularity of the site. Within the next few years, the prototype system had been converted into progressively improved versions, and these were substantially more effective than any other search engine then available.

As the buzz about their project spread, more and more people began to use it. Soon they were reporting that there were 10,000 searches per day at

[2]D. A. Vise and M. Malseed, *The Google Story*, Delacourt Press, New York, 2005.

[3]S. Brin and L. Page, *The Anatomy of a Large-Scale Hypertextual Web Search Engine*, Computer Science Department, Stanford University, Stanford, 1996.

[4]S. Brin and L. Page, "*The Future of the Internet*," speech to the Commonwealth Club, March 21, 2001.

Stanford on their system. With this growing use and popularity of their search system, they began to realize that they were maxing out their search ability due to the limited number of computers they had at their disposal. They would need more hardware to continue their remarkable expansion and enable more search activity. As Page said, "This is about how many searches we can do, and we need more computers. Our whole history has been like that. We always need more computers."[5]

In many ways, the research project at Stanford was a low budget operation. Because of a chronic shortage of cash, the pair are said to have monitored the Stanford Computer Science Department's loading docks for newly arrived computers to "borrow." In spite of this, within a short span of time, the reputation of the BackRub system had grown dramatically and their new search technology began to be broadly noticed.

They named their successor search engine *Google*, in a whimsical analogy to the mathematical term *googol*, which is the immensely large number 1 followed by 100 zeros. The transition from the earlier Backrub technology to the much more sophisticated Google was slow. But the Google system began with an index of 25 million pages and the capability to handle 10,000 search queries every day, even when it was in its initial stage of introduction. The Google search engine grew quickly as it was continuously improved. The effectiveness and relevance of the Google searches, its scope of coverage, its speed and reliability, and its clean user interface all contributed to a rapid increase in the popularity of the search engine.

At this time, Google was still a student research project, and both Page and Brin were still intent on completing their respective doctoral programs at Stanford. As a result, they initially refused to "answer the call" and continued to devote themselves to the academic pursuit of the technology of search.

Through all this, Brin maintained an eclectic collection of interests and activities. He continued with his graduate research interests at Stanford and he collaborated with his fellow Ph.D. students and professors on other projects such as automatic detection. At the same time, he also pursued a variety of outside interests, including sailing and trapeze. Brin's father had stressed the importance for him to complete his Ph.D. He said, "I expected him to get his Ph.D. and to become somebody, maybe a professor." In response to his father's question as to whether he was taking any advanced courses one semester, Brin replied, "Yes, advanced swimming."[6]

While Brin and Page continued on as graduate students, they began to realize the importance of what they had succeeded in developing. The two aspiring entrepreneurs decided to try and license the Google technology to existing Internet companies. But they found themselves unsuccessful in stimulating the interest of the major enterprises. They were forced to face the crucial decision of continuing on at Stanford or striking out on their own. With their

[5]D. A. Vise and M. Malseed, *The Google Story*, Delacourt Press, New York, 2005.
[6]Ibid.

realization that they were onto something that was important and perhaps even groundbreaking, they decided to make the move.

Thus our two heroes had reached their point of departure and they crossed over from the academic into the business world. As they committed to this new direction, they realized they would need to postpone their educational aspirations, prepare plans for their business concept, develop a working demo of their commercial search product, and seek funding sponsorship from outside investors.

Having made this decision, they managed to interest Sun Microsystems founder Andy Bechtolsheim in their idea. As Brin recalls, "We met him very early one morning on the porch of a Stanford faculty member's home in Palo Alto. We gave him a quick demo. He had to run off somewhere, so he said, 'Instead of us discussing all the details, why don't I just write you a check?' It was made out to Google Inc. and was for $100,000."[7]

The check remained in Page's desk uncashed for several weeks while he and Brin set up a corporation and sought additional money from family and friends—almost $1 million in total. Having started the new company, lined up investor funding, and possessing a superb product, they realized ultimate success would require a good balance of perspiration as well as inspiration. Nevertheless, at this point Google appeared to be well on the road to success.

Page and Brin have been on a roll every since, armed with the great confidence that they had both a superior product and an excellent vision for global information collection, storage, and retrieval. In addition, they believed that coordination and optimization of the entire hardware/software system was important, and so they developed their own Googleware technology by combining their custom software with appropriately integrated custom hardware, thereby fully leveraging their ingenious concept.

Google Inc. opened its doors as a business entity in September 1998, operating out of modest facilities in a Menlo Park, California garage.

As Page and Brin initiated their journey, they faced many challenges along the way. They matured in their understanding with the help of mentors they encountered such as Yahoo!'s Dave Filo. Filo not only encouraged the two in the development of their search technology but also made business suggestions for their project.

Following the company start-up, interest in Google grew rapidly. Red Hat, a Linux company, signed on as Google's first commercial customer. Red Hat was particularly interested in Google because the company realized the importance of search technology and its ability to run on open source systems such as Linux. In addition, the press began to take notice of this new commercial venture and articles began to appear in the media highlighting the Google product that offered relevant search results.

The late 1990s saw a spectacular growth in development of the technology industry, and Silicon Valley was awash with investor funding. The timing was

[7]"Search Us, Says Google," *Technology Review*, January 11, 2002.

right for Google, and in 1999, Page and Brin sought and received a second round of funding, obtaining $25 million from Silicon Valley venture capital firms.

The additional funding enabled them to expand their operations and move into new facilities they called the *Googleplex*, Google's current headquarters in Mountain View, California. Although at the time they occupied only a small portion of the new two-story building, they had clearly come a long way from a university research project to a full-fledged technology company with a rapid growth trajectory and a product that was in high demand.

Google was also in the process of developing a unique company culture. It operated in an informal atmosphere that facilitated both collegiality and an easy exchange of ideas. Google staffers enjoyed this rewarding atmosphere while they continued to make many incremental improvements to their search engine technology. For example, in an effort to expand the utility of their keyword-targeted advertising to small businesses, they rolled out the AdWords system, a software package that represents a self-service advertisement development capability.

Google took a major step forward when, in 2000, it was selected by Yahoo! to replace Inktomi as its provider of supplementary search results. Because of the superiority of Google over other search engine capabilities, licenses were obtained by many other companies, including the Internet services powerhouse America Online (AOL), Netscape, Freeserve, and eventually Microsoft Network (MSN). In fact, although Microsoft has pursued its own search technology, Bill Gates once commented on search engine technology development by saying that "Google kicked our butts."[8]

By the end of 2000, Google was handling more than 100 million searches each day. Shortly thereafter, Google began to deliver new innovations and establish new partnerships to enter the burgeoning field of mobile wireless computing. By expanding into this field, Google continued to pursue its strategy of putting search into the hands of as many users as possible.

As the global use of Google grew, the patterns contained within the records of search queries provided new information about what was on the minds of the global community of Internet users. Google was able to analyze the global traffic in Internet searching and identify patterns, trends, and surprises—a process they called *Google Zeitgeist*.

In 2004, Yahoo! decided to compete directly with Google and discontinued its reliance on the Google search technology. Nevertheless, Google continued to expand, increasing its market share and dominance of the Web search market through the deployment of regional versions of its software, incorporating language capabilities beyond English. As a result, Google continued to expand as a global Internet force.

Also in 2004, Google offered its stock to investors through an Initial Public Offering (IPO). This entrance into public trading of Google stock

[8]K. Kelleher, "Google vs. Gates," *Wired* 12(03): March 2004.

created not only a big stir in the financial markets but also great wealth for the two founding entrepreneurs. Page and Brin immediately joined the billionaire's club as they entered the exclusive ranks of the wealthiest people in the world.

Following the IPO, Google began to challenge Microsoft in its role as the leading provider of computer services. Google issued a series of new products, including the email service Gmail, the impressive map and satellite image product Google Earth, Google Talk to compete in the growing Voice over Internet Protocol (VoIP) market, and products aimed at leveraging the ambitious project to make the content of thousands of books searchable online, Google Base and Google Book Search. In addition to these new ventures, Google has continued to innovate in its core field of search by introducing new features for searching images, news articles, shopping services (Froogle), and other local search options.

It is clear that Google has become an essential tool for connecting people and information in support of the developing Information Revolution. Having established itself at the epicenter of the Web, Google is widely regarded as the "place to be" for the best and brightest programming talent in the industry. It is fair to say that, since the introduction of the printing press, no other entity or event has had more impact on public access to information than Google.

In fact, Google has endeavored to accumulate a good part of all human knowledge from the vast amount of information stored on the Web. The effective transformation of Google into an engine for what Page calls perfect search would basically give people everywhere the right answers to their questions and the ability to understand everything in the world.

Page and Brin could not have achieved their technological success without having a clear vision of the future of the Internet. Page recently commented in an interview that he believes that in the future "information access and communications will become truly ubiquitous," meaning that "anyone in the world will have access to any kind of information they want or be able to communicate with anyone else instantly and for very little cost." In fact, this vision of the future is not far from where we are now.[9]

Page also noted that the real power of the Internet is the ability to serve people all over the globe with access to information, which represents empowerment of individuals. The ability to facilitate the improved lives and productivity of billions of human beings throughout the world is an awesome potential outcome.

And the ability to support the information needs of people from different cultures and languages is an unusual challenge. Page stated in an interview that "even language is becoming less of a barrier. There's pretty good automatic translation out there. I've been using it quite a bit as Google becomes

[9]S. Brin and L. Page, *The Future of the Internet*, speech to the Commonwealth Club, March 21, 2001.

more globalized. It doesn't translate documents exactly, but it does a pretty good job and it's getting better every day."[10]

Even with translation and global reach, however, there remain significant challenges to connecting the people of the world through advanced information technology. One of the challenges is the potential for governmental restrictions on the access to information. Encryption technology, for example, inhibits the power of governments to monitor or control such information access. However, a 1998 survey of encryption policy found that several countries, including Belarus, China, Israel, Pakistan, Russia, and Singapore, maintained strong domestic controls while several other countries were considering the adoption of such controls.[11]

The phrase "Don't be evil" has been attributed to Google as its catch phrase or motto. Google's present CEO, Eric Schmidt, commented, in response to questions about the meaning of this motto, that "evil is whatever Sergey says is evil."

Brin, on the other hand, said in an interview with *Playboy Magazine*: "As for 'Don't be evil,' we have tried to define precisely what it means to be a force for good—always do the right, ethical thing. Ultimately 'Don't be evil' seems the easiest way to express it." And Page also commented on the phrase, saying "Apparently people like it better than 'Be good'."[12]

Page and Brin maintain lofty ambitions for the future of information technology, and they communicated those ambitions in an unprecedented seven-page letter to Wall Street entitled "An Owner's Manual for Google's Shareholders," written to detail Google's intentions as a public company. They explained their vision that "searching and organizing all the world's information is an unusually important task that should be carried out by a company that is trustworthy and interested in the public good."[13]

In response to questions about how Google will be used in the future, Brin said, "Your mind is tremendously efficient at weighing an enormous amount of information. We want to make smarter search engines that do a lot of the work for us. The smarter we can make the search engine, the better. Where will it lead? Who knows? But it's credible to imagine a leap as great as that from hunting through library stacks to a Google session, when we leap from today's search engines to having the entirety of the world's information as just one of our thoughts."[14]

At this junction, Page and Brin find themselves in a state of great personal wealth and great accomplishment, having created a technology and company that is profoundly affecting human culture and society. The two computer

[10]Ibid.

[11]"Cryptography and Liberty 1998, An International Survey of Encryption Policy," February 1998. http://www.gilc.org/crypto/crypto-survey.html.

[12]"Google Guys," *Playboy Magazine*, September 2004.

[13]From Google's letter to prospective shareholders. http://www.thestreet.com/_yahoo/markets/marketfeatures/10157519_6.html.

[14]"Google Guys," *Playboy Magazine*, September 2004.

scientists have traveled far in their hero's journey to carry out their vision of global search, having developed skills and capabilities for themselves as well as for Google and the Googleware technology. As they succeeded, their search technology became a key milestone in the development of the Information Revolution. Their journey is not over, however. Before continuing their story, let's digress into the historical context.

THE INFORMATION REVOLUTION

Over past millennia, the world has witnessed two global revolutions: the Agricultural Revolution and the Industrial Revolution.

During the Agricultural Revolution, a hunter-gatherer could acquire the resources from an area of 100 acres to produce an adequate food supply, whereas a single farmer needed only one acre of land to produce the equivalent amount of food. It was this 100-fold improvement in land management that fueled the Agricultural Revolution. It not only enabled far more efficient food production, but also provided food resources well above the needs of subsistence, resulting in a new era built on trade.

Where a single farmer and his horse had worked a farm, during the Industrial Revolution workers were able to use a single steam engine that produced 100 times the horsepower of this farmer–horse team. As a result, the Industrial Revolution placed a 100-fold increase of mechanical power into the hands of the laborer. It resulted in the falling cost of labor and this fueled the unprecedented acceleration in economic growth that ensued.

Over the millennia, humans have accumulated great knowledge, produced a treasury of cultural literature, and developed a wealth of technology advances, much of which has been recorded in written form. By the mid-20th century, the quantity of accessible useful information had grown explosively, requiring new methods of information management; and this can be said to have triggered the Information Revolution. As computer technology offered great improvements in information management technology, it also provided substantial reductions in the cost of information access. It did more than allow people to receive information. Individuals could buy, sell, and even create their own information. Cheap, plentiful, easily accessible information has become as powerful an economic dynamic as land and energy had for the two prior revolutions.

The falling cost of information has, in part, reflected the dramatic improvement in price–performance of microprocessors, which appears to be on a pattern of doubling every eighteen months. While the computer has been contributing to information productivity since the 1950s, the resulting global economic productivity gains were initially slow to be realized.

Until the late 1990s, networks were rigid and closed, and time to implement changes in the telecommunication industry was measured in decades. Since then, the Web has become the grim reaper of information inefficiency.

For the first time, ordinary people had real power over information production and dissemination. As the cost of information dropped, the microprocessor in effect gave ordinary people control over information about consumer products.

Today, we are beginning to see dramatic change as service workers experience the productivity gains from rapid communications and automated business and knowledge transactions. A service worker can now complete knowledge transactions 100 times faster using intelligent software and near ubiquitous computing in comparison to a clerk using written records. As a result, the Information Revolution is placing a 100-fold increase in transaction speed into the hands of the service worker. Therefore, the Information Revolution is based on the falling cost of information-based transactions, which in turn fuels economic growth.

In considering these three major revolutions in human society, a defining feature of each has been the requirement for more knowledgeable and more highly skilled workers. The Information Revolution signals that this will be a major priority for its continued growth. Clearly, the Web will play a central role in the efficient performance of the Information Revolution because it offers a powerful communication medium that is itself becoming ever more useful through intelligent applications.

Over the past fifty years, the Internet/World Wide Web has grown into the global Information Superhighway. And just as roads connected the traders of the Agricultural Revolution and railroads connected the producers and consumers of the Industrial Revolution, the Web is now connecting people to information in the Information Revolution.

The Information Revolution enables service workers today to complete knowledge transactions many times faster through intelligent software using photons over the Internet, in comparison to clerks using electrons over wired circuits just a few decades ago.

But perhaps the most essential ingredient in the Web's continued success has been search technology such as Google, which has provided real efficiency in connecting to relevant information and completing vital transactions. Now Google transforms data and information into useful knowledge, energizing the Information Revolution.

DEFINING INFORMATION

Google started with Page and Brin's quest to mine data and make sense of the voluminous information on the Web. But what differentiates information from knowledge and how do companies like Google manipulate it on the Web to nourish the Information Revolution?

First, let's be clear about what we mean by the fundamental terms *data, information, knowledge*, and *understanding*.

An item of data is a fundamental element of information, the processed data that has some independent usefulness. And right now data is the main thing you can find directly on the Web in its current state. Data can be considered the raw material of information. Symbols and numbers are forms of data.

Data can be organized within a database to form structured information. While spreadsheets are "number crunchers," databases are the "information crunchers." Databases are highly effective in managing and manipulating structured data.[15]

Consider, for example, a directory or phone book that contains elements of information (i.e., names, addresses, and phone numbers) about telephone customers in a particular area. In such a directory, each customer's information is laid out in the same pattern. The phone book is basically a table that contains a record for each customer. Each customer's record includes his/her name, address, and phone number. But you can't directly search such a database on the Web. This is because there is no schema defining the structure of data on the Web. Thus what looks like information to the human being who is looking at the directory (taking with him his background knowledge and experience as a context) in reality is data because it lacks this schema.

On the other hand, information explicitly associates one set of things to another. A telephone book full of data becomes information when we associate the data to persons we know or wish to communicate with.

For example, suppose we found data entries in a telephone book for four different persons named Jones, but all of them were living within one block of each other. The fact that there are four bits of data about persons with the same name in approximately the same location is interesting information.

Knowledge, on the other hand, can be considered to be a meaningful collection of useful information. We can construct information from data. And we can construct knowledge from information. Finally, we can achieve understanding from the knowledge we have gathered.

Understanding lies at the highest level. It is the process by which we can take existing knowledge and synthesize new knowledge. Once we have understanding, we can pursue useful actions because we can synthesize new knowledge or information from what is previously known.

Again, knowledge and understanding are currently elusive on the Web. Future Semantic Web architectures seek to redress this limit.

To continue our telephone example, suppose we developed a genealogy tree for the Joneses and found the four Joneses who lived near each other were actually brothers. This would give us additional knowledge about the Joneses in addition to information about their addresses. If we then interviewed the brothers and found that their father had bought each brother a house in his neighborhood when they married, we would finally understand

[15] "Databases from Scratch I: Introduction." http://brebru.com/databases_from_scratch_1.html.

quite a bit about them. We could continue the interviews to find out about their future plans for their offspring—thus producing more new knowledge.

If we could manipulate data, information, knowledge, and understanding by combining a search engine, such as Google, with a reasoning engine, we could create a logic machine. Such an effort would be central to the development of artificial intelligence (AI) on the Web.

AI systems seek to create understanding through their ability to integrate information and synthesize new knowledge from previously stored information and knowledge. An important element of AI is the principle that intelligent behavior can be achieved through processing of symbolic structures representing increments of knowledge. This has produced knowledge-representation languages that allow the representation and manipulation of knowledge to deduce new facts from the existing knowledge.

The World Wide Web has become the greatest repository of information on virtually every topic. Its biggest problem, however, is the classic problem of finding a needle in a haystack. Given the vast stores of information on the Web, finding exactly what you're looking for can be a major challenge. This is where search engines, like Google, come in—and where we can look for the greatest future innovations to come when we combine AI and search.

Larry Page and Sergey Brin found that the existing search technology looked at information on the Web in simple ways. They decided that to deliver better results, they would have to go beyond simply looking, to looking good.

LOOKING GOOD

Commercial search engines are based on one of two forms of Web search technologies: human-directed search and automated search. Human-directed search is search in which the human performs an integral part of the process. In this form of search engine technology, a database is prepared of keywords, concepts, and references that can be useful to the human operator. Searches that are keyword based are easy to conduct but they have the disadvantage of providing large volumes of irrelevant or meaningless results. The basic idea in its simplest form is to count the number of words in the search query that match words in the keyword index and rank the Web page accordingly. Although more sophisticated approaches also take into account the location of the keywords, the improved performance may not be substantial. As an example, it is known that keywords used in the title tags of Web pages tend to be more significant than words that occur in the Web page, but not in the title tag; however, the level of improvement may be modest.

Another approach is to use hierarchies of topics to assist in human-directed search. The disadvantage of this approach is that the topic hierarchies must be independently created and are therefore expensive to create and maintain.

The alternative approach is automated search; this approach is the path taken by Google. It uses software agents, called Web crawlers (also called spiders, robots, bots, or agents), to automatically follow hypertext links from one site to another on the Web until they accumulate vast amounts of information about the Web pages and their interconnections. From this, a complex index can be prepared to store the relevant information. Such automated search methods accumulate information automatically and allow for continuing updates.

However, even though these processes may be highly sophisticated and automatic, the information they produce is represented as links to words, and not as meaningful concepts.

Current automated search engines must maintain huge databases of Web page references. There are two implementations of such search engines: individual search engines and meta-searchers. Individual search engines (such as Google) accumulate their own databases of information about Web pages and their interconnections and store them in such a way as to be searchable. Meta-searchers, on the other hand, access multiple individual engines simultaneously, searching their databases.

In the use of keywords in search engines, there are two language-based phenomena that can significantly impact effectiveness and therefore must be taken into account. The first of these is polysemy, the fact that single words frequently have multiple meanings; and the second is synonymy, the fact that multiple words can have the same meaning or refer to the same concept.

In addition, there are several characteristics required to improve a search engine's performance. It is important to consider useful searches as distinct from fruitless ones. To be useful, there are three necessary criteria: (1) maximize the relevant information, (2) minimize irrelevant information, and (3) make the ranking meaningful, with the most highly relevant results first.

The first criterion is called recall. The desire to obtain relevant results is very important, and the fact is that, without effective recall, we may be swamped with less relevant information and may, in fact, not receive the most important and relevant results. It is essential to reduce the rate of false negatives—important and relevant results that are not displayed—to a level that is as low as possible.

The second criterion, minimizing irrelevant information, is also very important to ensure that relevant results are not swamped; this criterion is called precision. If the level of precision is too low, the useful results will be highly diluted by the uninteresting results, and the user will be burdened by the task of sifting through all of the results to find the needle in the haystack. High precision means a very low rate of false positives, irrelevant results that are highly ranked and displayed at the top of our search result.

Since there is always a trade-off between reducing the risk of missing relevant results and reducing the level of irrelevant results, the third criterion, ranking, is very important. Ranking is most effective when it matches our information needs in terms of our perception of what is most relevant in our results. The challenge for a software system is to be able to accurately match the expectations of a human user since the degree of relevance of a search contains several subjective factors such as the immediate needs of the user and the context of the search. Many of the desired characteristics for advanced search, therefore, match well with the research directions in artificial intelligence and pattern recognition. By obtaining an awareness of individual preferences, for example, a search engine could more effectively take them into account in improving the effectiveness of search.

Recognizing ranking algorithms were the weak point in competing search technology, Page and Brin introduced their own new ranking algorithm—PageRank.

GOOGLE CONNECTS INFORMATION

Just as the name Google is derived from the esoteric mathematical term googol, in the future, the direction of Google will focus on developing the esoteric "perfect search engine," defined by Page as something that "understands exactly what you mean and gives you back exactly what you want." In the past, Google has applied great innovation to try and overcome the limitations of prior search approaches; PageRank was conceived by Google to overcome some of the key limitations.[16]

Page and Brin recognized that providing the fastest, most accurate search results would require a new approach to server systems. While most search engines used a small number of large servers that often slowed down under peak use, Google went the other direction by using large numbers of linked PCs to find search results in response to queries. The approach turned out to be effective in that it produced much faster response times and greater scalability while minimizing costs. Others have followed Google's lead in this innovation while Google has continued its efforts to make its systems more efficient.

Google takes a parallel processing approach to its search technology by conducting a series of calculations on multiple processors. This has provided Google with critical timing advantage, permitting its search algorithms to be very fast. While other search engines rely heavily on the simple approach of counting the occurrences of keywords, Google's PageRank approach considers the entire link structure of the Web to help in the determination of Web page importance. By then performing a hypertext matching assessment to narrow the search results for the particular search being conducted, Google achieves superior performance. In a sense, Google combines insight into Web page importance with query-specific attributes to rank pages and deliver the most relevant results at the top of the search results.

The PageRank algorithm analyzes the importance of the Web pages it considers by solving an exceptionally complex set of equations with a huge

[16]Quotes from `http://www.google.com/corporate/tech.html`.

number of variables and terms. By considering links between Web pages as "votes" from one page to another, PageRank can assign a measure of a page's importance by counting its votes.

It also takes into account the importance of each page that supplies a vote and, by appropriately weighting these votes, further improves the quality of the search. In addition, PageRank considers the Web page content, but unlike other search engines that restrict such consideration to the text content, Google consider the full contents of the page.

In a sense, Google attempts to use the collective intelligence of the Web, a topic for further discussion later in this book, in its effort to improve the relevance of its search results. Finally, because the search algorithms used by Google are automated, Google has earned a reputation for objectivity and lack of bias in its results.

Throughout their exciting years establishing and growing Google as a company, Page and Brin realized that continued innovation was essential. They undertook to find innovative services that would enhance access to Web information with added thought and not a little perspiration. Page said that he respected the idea of having "a healthy disregard for the impossible."[17]

In February 2002, the Google Search Appliance, a plug-and-play application for search, was introduced. In short order, this product was dispersed throughout the world populating company networks, university systems, and the entire Web. The popular Google Search Appliance is referred to as "Google in a box."

In another initiative, Google News was introduced in September 2002. This free news service, which allows automatic selection and arrangement of news headlines and pictures, features real-time updating and tailoring, allowing users to browse the news with scan and search capabilities.

Continuing Google's emphasis on innovation, the Google search service for products, Froogle, was launched in December 2002. Froogle allows users to search millions of commercial websites to find product and pricing information. It enables users to identify and link to a variety of sources for specific products, providing images, specifications, and pricing information for the items being sought.

Google's innovations have also impacted the publishing business with both search and advertising features. Google purchased Pyra Labs in 2003 and thus became the host of Blogger, a leading service for the sharing of thoughts and opinions through online journals, or blogs (weblogs).

Finally, Google Maps became a dynamic online mapping feature, and Google Earth a highly popular mapping and satellite imagery resource. Using these innovative applications, users can find information about particular locations, get directions, and display both maps as well as satellite images of a desired address.

[17]D. A. Vise and M. Malseed, *The Google Story*, Delacourt Press, New York, 2005.

With each new capability, Google expands our access to more information[18] and moves us closer to Page's holy grail—perfect search.

At this junction, Page and Brin have finally completed their hero's journey. They have become the *masters of search*; committed to improving access to information and lifting the bonds of ignorance from millions around the world.

PATTERNS OF DISCOVERY

Larry Page and Sergey Brin were trying to solve the problem of easy, quick access to all Web information, and ultimately to all human knowledge. In order to index existing Web information and provide rapid relevant search results, their challenge was to sort through billions of pages of material efficiently and explicitly find the right responses.

They were confident that their vision for developing a global information collection, storage, and retrieval system would succeed if they could base it on a unique and efficient ranking algorithm.

The process of inspiration for Page and Brin became fulfilled when they completed their seminal paper entitled *The Anatomy of a Large-Scale Hypertextual Web Search Engine*, which explained their efficient ranking algorithm, PageRank. In developing a breakthrough ranking algorithm based on the ideas of publication ranking, Page and Brin experienced a moment of inspiration.

But they didn't stop there. They also believed that optimization was vitally important and so they developed their own Googleware technology consisting of combining custom software with custom hardware, thereby reflecting the founders' genius. They built the world's most powerful computational enterprise, and they have been on a roll ever since. Page stressed that inspiration still required perspiration and that Google appeared destined for rapid growth and expansion. In building the customized computer Googleware infrastructure for PageRank, they were demonstrating the 1% Inspiration and 99% Perspiration Pattern.

The result was Google, the dominant search engine connecting people to all of the World Wide Web's information.

FORECAST FOR CONNECTING INFORMATION

For many of us it seems that an uncertain future looms ahead like a massive opaque block of granite. But just as Michelangelo suggested that he took a block of stone and chipped away the nonessential pieces to produce *David*, we can chip away the improbable to uncover the possible. By examining inventors and their process of discovery, we are able to visualize the tapestry of our past to help unveil patterns that can serve as our guideposts on our path forward.

[18]S. Brin and L. Page, "The Anatomy of a Large-Scale Hypertexual wab Search Engine," *Computer Networks*, 30(1–7), 107–117, 1998.

Page and Brin invented an essential search technology, but their contributions to information processing were evolutionary in nature—built on inspiration and perspiration. One forecast for connecting information is that we can expect a continued pattern of inspired innovation as we go forward in the expansion of search and related technology.

Discoveries Requiring Inspiration and Perspiration

In considering the future for connecting information, we expect that improved ranking algorithms will ensure Google's continued dominance for some time to come. Extrapolating from Google's success, we can expect a series of inspired innovations building on its enterprise computer system, such as offering additional knowledge-related services.

Future Google services could include expanding into multimedia areas such as television, movies, and music using Google TV and Google Mobile. Viewers would have all the history of TV from which to choose. And Google would offer advertisers targeted search. Google Mobile could deliver the same service and products to cell phone technology. By 2020, Google could digitize and index every book, movie, TV show, and song ever produced, making it available conveniently.

In addition, Google could dominate the Internet as a hub site. The ubiquitous GoogleNet would dominate wireless access and cell phones. As for the Google browser, Gbrowser, it could replace operating systems.

However, our vision also includes connecting information through developing more intelligent search capabilities. A new Web architecture, such as Tim Berners-Lee's Semantic Web, would add knowledge representation and logic to the markup languages of the Web. Semantics on the Web would offer extraordinary leaps in Web search capabilities.

Since Google has cornered online advertising, it has become progressively more precision targeted and inexpensive. But Google also has 150,000 servers with nearly unlimited storage space and massive processing power.

Beyond simply inspired discoveries, Google or other search engine powers could find innovations based on new principles yet to be proved, as suggested in the following.

Discoveries Requiring New Proof of Principle

Technology futurists such as Ray Kurzweil have suggested that Strong AI (software programs that exhibit true intelligence) could emerge from developing Web-based systems such as that of Google. Strong AI could perform data mining at a whole new level. This type of innovation would require a proof of principle.

Some have suggested that Google's purpose in converting books into electronic form is not to provide for humans to read them, but rather to provide a form that could be accessible by software, with AI as the consumer.

One of the great areas of innovation resulting from Google's initiatives is its ability to search the human genome. Such technology could lead to a personal DNA search capability within the next decade. This could result in the identification of medical prescriptions that are specific to you; and you would know exactly what kinds of side effects to expect from a given drug.

And consider what might happen if we had the perfect search. Think about the capability to ask any question and get the perfect answer—an answer with real context. The answer could incorporate all of the world's knowledge using text, video, or audio. And it would reflect every nuance of meaning. Most importantly, it would be tailored to your own particular context. That's the stated goal of IBM, Microsoft, Google, and others. Such a capability would offer its greatest benefits when knowledge is easily gathered.

Soon search will move away from the PC-centric operations to the Web connected to many small devices such as mobile phones and PDAs. The most insignificant object with a chip and the ability to connect will be network aware and searchable. And search needs to solve access to deep databases of knowledge, such as the University of California's library system. While there are several hundred thousand books online, there are 100 million more that are not.

The perfect search will find all this information and connect us to the world's knowledge, but this is the beginning of decision making, not the end. Search and artificial intelligence seem destined to get together.

In the coming chapters, we will be exploring all the different technologies involved in connecting information and we will be exploring how the prospects for the *perfect search* could turn into *ubiquitous intelligence*.

First, ubiquitous computing populates the world with devices using microchips everywhere. Then the ubiquitous Web connects and controls these devices on a global scale. The ubiquitous Web is a pervasive Web infrastructure that allows all physical objects access by URIs, providing information and services that enrich users' experiences in their physical context just as the Web does in cyberspace. The final step comes when artificial intelligence reaches the capability of managing and regulating devices seamlessly and invisibly within the environment—achieving ubiquitous intelligence.

Ubiquitous intelligence is the final step of Larry Page's perfect search and the future of the Information Revolution.

2

Connecting Circuits

When I am working on a problem I never think about beauty.
I only think about how to solve the problem. But when I have finished,
if the solution is not beautiful, I know it is wrong.
—Buckminster Fuller[1]

The mighty Mississippi River begins as a stream in northern Minnesota. At that point, it is about 10 meters wide and less than 1 meter deep. As it makes its way south to the Gulf of Mexico, it is fed by its tributaries—the Missouri, Illinois, Ohio, Arkansas, and Red rivers—and grows until it reaches New Orleans, where it is nearly 70 meters deep and 2200 meters across. As a result, New Orleans has become the largest commercial seaport in the country, transporting the unending flow of goods that helped create the Industrial Revolution.

In a similar fashion, the Information Superhighway began as embryonic inventions including the vacuum tube, transistor, and microchip—each a "tributary" invention feeding into the evolution of the computer. At that point computing consisted of isolated machines as big as a room. With the miniaturization of microprocessors following Moore's Law, networks of computers grew into the Internet. Bandwidth and access expanded to deliver data globally. Now the Information Superhighway has become a spectacular flood, transporting an unending flow of data to create the Information Revolution.

Moore's Law is commonly represented as the observation that the cost–performance of microelectronic components doubles every 18 months. It is a centerpiece of the computer's evolution. In this chapter, we present the story

[1]B. Fuller, "Quote DB." `http://www.quotedb.com/quotes/107`.

Connections: Patterns of Discovery By H. Peter Alesso and Craig F. Smith
Copyright © 2008 John Wiley & Sons, Inc.

of Moore's Law and its impact on our rapid transformation from an industrial society into an information society. As part of the story, we consider the inventors Thomas Edison, John Fleming, Lee de Forest, John Mauchly, J. Presper Eckert, John Bardeen, Walter Brattain, William Shockley, Robert Noyce, Jack Kilby, and Gordon Moore as they helped to connect circuits and opened the floodgates to information processing. We follow their contributions to the tributary inventions—the vacuum tube, transistor, and microchip—that feed into the development of the computer—creating Moore's Law and the global "river" of data on the Information Superhighway.

MOORE'S LAW STORY

There can be enormous power in tiny things—consider the microchip. Smaller than a penny, the chip, or integrated circuit (IC), is literally the brain and nervous system of every digital device in the world. By *connecting circuits*, chips enable computers, mobile phones, automobiles, and satellites. Chips are used to design space stations and football stadiums. They deliver hundreds of TV channels as well as the Internet directly to the consumer. Since the commercial introduction of the microchip in the early 1970s, scientific breakthroughs have been made in medicine, mathematics, and engineering at an ever accelerating pace.

The computer chip is a central character in the discussion of connecting circuits because it is the centerpiece technology that has been fueling the Information Revolution; it is the continuing and ever accelerating pace of growth of chip capacity that has been the driving force behind innovation. Information technology is estimated to grow at an exponential rate because this tiny chip seems to follow an ambiguous rule called Moore's Law.

Moore's Law takes its name from Gordon Moore who, in 1965, wrote an article[2] in *Electronics Magazine* in which he contemplated the future of the semiconductor industry. Moore, who shared in the invention of the microprocessor chip and went on to cofound Intel Corporation, noted that the density of components on semiconductor chips had doubled yearly since the initial prototypes were introduced in 1959. This annual doubling in component density, amounting to an exponential growth rate, soon became widely known as Moore's Law.

From the beginning, it has been recognized that Moore's Law is not a scientific or physical law in the sense of Newton's law of gravitation, which is a statement of how the world works. Rather, it is the result of the observation of a continuing exponential growth of circuit density, driven by a series of technology advances developed in response to continued and robust commercial demand for ever smaller electronic devices. Moore made the purely

[2]G. Moore, "Cramming More Components onto Integrated Circuits," *Electronics Magazine* 38(8): April 19, 1965.

empirical observation that the increases in circuit density appeared to be continuing with a one-year doubling time.

Although Moore's Law was initially considered to be well represented by an annual or 12-month doubling rate, by the time the 1990s rolled around, more data had accumulated, and it was possible to determine a more accurate estimate of past and continuing growth in the density of transistors on semiconductor chips. An 18-month doubling time was seen to be a much better reflection of the actual and projected growth rates. And so, it became commonly accepted that Moore's Law reflected the empirical observation of an 18-month doubling time in the complexity of semiconductor chips and, perhaps more importantly, of the computing power represented by a given cost.

Moore not only observed the past rate of growth in chip capacity, but he also used the observed growth rates to project future growth. It is common now for technology planners and prognosticators to use Moore's Law to predict continuing growth in chip capacity, with a doubling time of 18 months. The Moore's Law observation has seen the capacity of memory chips rise from one thousand bits in 1971 to one million bits in 1991 and to one billion bits by 2001. The billion-bit semiconductor memory chip represents an extraordinary growth in capacity of nine orders of magnitude, and a similar growth rate has also been experienced in the capability of microprocessor chips to process data.

The microchip is a marvel of connected circuits that amplify current and turn switches on and off. Just as the origins of the Mississippi can be traced to a stream in northern Minnesota, the origin of microchips can be traced to the workbench of Thomas Edison at Menlo Park, New Jersey.

EDISON'S ELECTRIC LIGHT

In 1878, Edison became intrigued with the challenge of electric lighting. He arrived on the scene just as the world was ready for such a technology. Proof of principle experiments had already produced light from heat. Edison took a 1% Inspiration and 99% Perspiration approach to the problem. With his inspired insight, he saw that a filament of high electrical resistance, in a parallel circuit protected in a sealed and evacuated bulb, could produce a lasting incandescent light.

In 1879, Edison was able to convert this insight into his invention of the light bulb—a practical and reliable electric light—by assembling a carbonized filament in a transparent bulb with a relatively high vacuum (to protect the filament from burning) and energizing it with electricity at a low current flow. As important as the basic invention of the electric light was, Edison's key achievement was not just the incandescent electric light, but the complete electric lighting system. The lighting system Edison introduced consisted of all the components needed to make the electrical light practical, safe, and economical, including not only the light bulb but also electricity generating equipment (i.e., the dynamo); a network of wire conductors for power distribution; safety materials and devices such as insulators and fuses; switches; voltage control equipment; and so on.

Today, the electric power utility industry represents a major component of the industrial economies of the modern world. This industry had its modest start when, in New York City, the first commercial power station began operations in 1882 for the purpose of providing electricity for public lighting and residential use in a small area of Manhattan. This new industry provided for reliable central distribution of electricity that, in combination with the newly invented practical light bulb, quickly became competitive with gas lighting.

The development of the electric lighting system followed an S-shaped curve of growth. In this process, typical for new technology introduction, the new concept for electric light was first introduced as a novelty and it experienced a low level of initial adoption; this was followed by the rapid growth that resulted when electric lighting replaced gas lighting systems while expanding the overall market; finally, growth reached a relative plateau as saturation of the market was reached.

The invention of the electric lighting system, by itself, was an impressive demonstration of the creativity and innovation of Edison. However, there was an unexpected side effect that occurred as a by-product of the invention of the light bulb that was to have a profound impact on the future of the Information Revolution—this was the discovery of a vacuum tube phenomenon known as the Edison Effect.

THE VACUUM TUBE DIODE

Much was to come of Edison's light bulb consisting of a filament mounted in a glass bulb with a vacuum. In this simple (though challenging to accomplish) concept, the electricity that travels through the filament results in its being heated to a sufficiently high temperature that it glows with heat, or incandesces, and then radiates light. The evacuation of the bulb allows the filament to survive high temperatures without burning or oxidizing.

Edison experimented with his light bulb and in 1883 he found that he could detect electrons flowing through the vacuum to a metal plate attached inside the bulb. This was a serendipitous discovery by Edison who did not set out to invent this process. The phenomenon subsequently became widely known as the Edison Effect.

Later, while John Fleming, a British physicist, was investigating the Edison Effect, he discovered it could be used to detect radio waves and to convert them into electricity. Fleming created a variation on the Edison light bulb to investigate this phenomenon further.

Fleming found that his device was capable of converting an alternating current signal into direct current. This device, known as a *diode*, was little more than an Edison light bulb with an extra electrode mounted inside. When the diode's filament is heated to a white-hot temperature, electrons are released as they are boiled off the surface of the filament. By making the added electrode (called the *plate* or *anode*) more positive in electrical potential than the hot filament, a current is created that flows through the vacuum. And since the filament is much hotter than the extra electrode, the current tends to flow only in one direction: from the filament to the electrode. As a result, the arrangement in the bulb allows alternating current signals to be converted into direct current flow. Fleming's diode was first used as a sensor of the weak signals produced by the new wireless telegraph. Later, in addition to many other functions, the diode vacuum tube was used to convert AC into DC in power supplies for electronic equipment.[3]

Fleming's work is an example of the second type of pattern of discovery. Here Fleming used a known phenomenon—the Edison Effect—to invent a proof of principle application—the vacuum tube diode.

Other inventors were inspired by Fleming's work and attempted to improve on it. Lee de Forest, in particular, succeeded where others failed. De Forest was another prolific American inventor of the time. By adding a third electrode, called the grid, into the diode tube, he found that the resulting *triode* had unique characteristics that allowed it to be used as an on–off switch as well as an amplifier. The triode found great application in the technology for radio transmitters and elsewhere.

In 1907, de Forest patented his new device. The added electrode, or *grid* electrode, was simply a bent wire positioned between the plate and the filament. The additional electrode changed the characteristics of the tube in an important way. If a signal from a wireless telegraph antenna was applied to the grid instead of the filament, it was possible to obtain a much more sensitive detection of a weak signal. In fact, the grid was capable of changing, or modulating, the current flowing from the filament to the plate. The resulting device, known as the Audion, was the first successful electric signal amplifier. It was a key element in the genesis of today's electronics industry. The resulting vacuum tube technology was to play a particularly important role in the development of calculating machines.[4]

The work of de Forest in inventing the Audion illustrates a form of the third type of pattern of discovery. Building on the phenomenon of the Edison Effect and the proof of principle of Fleming's diode, de Forest was able to persevere to invent a commercially important vacuum tube that would be the key to calculating machines—the electronic amplifier.

The Audion provided the basis for all subsequent vacuum tube technology. It consists of a heated *cathode* boiling off electrons into a vacuum. The electrons that are emitted from the cathode travel through one or more grids, and this allows control of the electron current. When the electrons ultimately strike the anode (plate), they are absorbed there, completing the circuit. By selecting

[3]"Vacuum Tube Valley." August 8, 2001. `http://www.vacuumtube.com/toppage11.htm`.
[4]Ibid.

the dimensions and materials of these tube components properly, it is possible to take a small AC signal and increase the voltage of the signal, thereby amplifying it.[5]

Over two decades passed from the time of the original discovery of the phenomenon called the Edison Effect until practical applications of the diode and Audion were developed. And several more decades passed until these fundamental building blocks of vacuum tube technology became available for computer applications.

From the early 1900s to the mid-1960s, an astounding variety of different vacuum tubes were designed, fabricated, and sold. Application of vacuum tube technology found its way into virtually all forms of electronic equipment—from televisions and radios to electric organs and radar systems. One of the most important of the applications of vacuum tube technology, however, was the electronic computing machine.

Of all of the features of vacuum tubes, perhaps the most important one, and the one that is most relevant to our story of Moore's Law, is its ability to perform the function of an ON–OFF switch, and it is this feature that found its application in computer technology.

It was also the ability of vacuum tubes to serve as ON–OFF switches combined with the possibility of interconnecting large numbers of these specialized electronic devices that inspired the development of the computer during World War II.

THE FIRST PROGRAMMABLE COMPUTERS

During World War II, secrecy in communications was exceptionally important as both sides attempted to conduct global military operations directed from remote headquarters. To ensure secrecy in their communications, the German military developed an encryption device called the *Enigma* machine. Enigma was an advanced coding device that could encrypt information with a huge number of different coding patterns. The first Enigma machines used three rotors, each of which could be separately set, and were able to encrypt messages with more than 8 million code possibilities. Subsequently, more advanced Enigma machines were used, and the number of code possibilities became unimaginably large. The Germans believed that messages sent with the Enigma system were completely secure since the code was virtually unbreakable. The drive to break the German codes would turn out to be a major impetus to the further development of electronic computing machines.

Although the Germans believed their Enigma coding system to be unbreakable, the Allies succeeded in breaking the codes without the Germans being aware of their success. The Allies broke the code with a new code decryption system, called the *Bombe*. The Bombe was based on prewar code-breaking

[5]Ibid.

efforts in Poland that resulted in a machine they called the *Bomba*. The Poles had shared their technology with their British counterparts before the war, and once the war began, the British devoted considerable effort to breaking the German codes using the Bombe electromechanical decryption technology.

As the war progressed, the Germans continued to improve their encryption machines, eventually developing a system that included ten rotors, which operated in an irregular fashion and which finally surpassed the capability of the British Bombe technology. When the British realized that a new approach was needed, the brilliant scientists and engineers at Britain's establishment for code breaking, Bletchley Park, and elsewhere, conceived of, designed, and built the world's first completely electronic computer, called *Colossus*. Like the Bombe, Colossus was also successful in secretly breaking the German codes. Both systems were of great importance in the successful war effort of the Allied Forces.

Another very important early developmental effort was the completion of the Harvard Mk I computer. Built by IBM at Harvard University in response to the U.S. military's desire to automate tedious computations such as the calculation of firing tables for battlefield artillery, the Harvard Mark I was the first programmable electronic computer in the United States.

Thus, the origin of the technology of modern computing can be traced directly to two World War II era military-related projects, Colossus and the Harvard Mk I. These two pioneering computer systems can be considered to be the first of the *First Generation* computers. First Generation computers are systems that are based on wired circuits, vacuum tubes, and punched cards.

ENIAC

John Mauchly and J. Presper Eckert, Jr. are two American computer pioneers and innovators who very early on considered the possibility of building a large, general purpose electronic computer. The computer would be based on the use of vacuum tube technology to serve the function of switches, each of which could be placed in one of two binary states: 5 volts, representing the ON state, and less than 5 volts, representing the OFF state. In 1942 Mauchly outlined his ideas for building the first large-scale electronic digital computer whose purpose would be to perform general numerical computations. Mauchly was a former student of the much younger Eckert, and the proposed initiative captured the attention of both men who entered into a highly important collaboration.

John Mauchly was born in 1907 and he grew up living in the Washington DC area. His father was a physicist, and John, from an early age, exhibited interest in technology. He was awarded a scholarship to study engineering at Johns Hopkins University in Baltimore. Although he started out studying engineering, he eventually moved toward his academic interests in pure science and completed his undergraduate studies with a degree in physics. He continued on in his study of physics, and after he earned his Ph.D. in 1932, he went on to teach physics. From his interest in weather prediction, he realized early on that the ability to automate complex calculations would have great value.

J. Presper Eckert, Jr. was born in 1919, the only child of a wealthy Philadelphia businessman and his wife. Interested in technology from an early age, he had wanted to attend MIT, but because of his parents' desire to have him stay closer to home, he enrolled in the Moore School of Electrical Engineering at the University of Pennsylvania from which he received his bachelor's degree in 1941 and his master's degree in 1943. A brilliant student, he was given a position as an instructor at the Moore School immediately upon his graduation. While Eckert was an instructor at the Moore School, the two pioneers' paths crossed when Mauchly attended a course there which Eckert was teaching.

Following extensive discussions between the two on the development of a large-scale general purpose computer, and Mauchly's written description of his plan, the two decided to work together on the keystone project that was to be called the Electronic Numerical Integrator and Calculator (ENIAC).

ENIAC was to be the first large-scale digital computer capable of general purpose programming. It was funded by the U.S. Army and was built by a team led by Mauchly and Eckert at the University of Pennsylvania. The ENIAC design called for more than 100,000 separate components, and the resulting machine was 100 feet by 10 feet by 3 feet in dimension. Although it suffered from poor reliability, it provided an acceptable solution to a problem that had vexed scientists and engineers for centuries: how to avoid the time and drudgery of complex repetitive arithmetic calculations.

Like Colossus and the Harvard Mk I, Mauchly and Eckert's ENIAC was also a First Generation machine, being constructed of 17,468 vacuum tubes and wire conductors with punched cards for input, output, and storage. It was unveiled February 14, 1946. Among its early missions was its use by the military to perform calculations related to ballistic trajectories. In short order it was used for such diverse applications as the design of the hydrogen bomb, weather prediction, cosmic ray studies, and wind tunnel design. At the time, it was by far the largest single electronic apparatus in the world. Weighing in at over 30 tons and requiring 200 kilowatts of electrical power, ENIAC was an impressive industrial scale machine. ENIAC could perform 5000 additions/subtractions and 300 multiplications per second, paltry performance from today's perspective, but at the time, 1000 times faster than any contemporary machine. Most importantly, ENIAC represents the beginning of the era of modern computer technology.

After the completion of ENIAC, Mauchly and Eckert left the University of Pennsylvania to set up a new enterprise called the Electronic Controls Company. (They subsequently changed the name of the company to the Eckert–Mauchly Computer Corporation (EMCC) in December 1947.) Eckert began the development of a new computer system and Mauchly focused on strategic research into future uses of automatic data processing systems. The new company began designing electronic computing systems for their first clients, the BINAC (Binary Automatic Computer) for Northrop Aircraft Company and then the UNIVAC-I (Universal Automatic Computer) for the U.S. Census Bureau.

In early 1949, EMCC started up the BINAC, which used magnetic tape to store data and had two independent central processing units (CPUs). While being tested at the EMCC facilities, the system performed well and passed its operating tests. Later in 1949, BINAC was shipped to Northrop. The performance of BINAC was somewhat disappointing to Northrop, possibly due to damage in shipping.

In early 1950, Eckert and Mauchly sold their company to Remington-Rand, where it became an operating division known as the Univac Division. Their work for the Census Bureau was completed when they delivered the UNIVAC-I to the Census Bureau in June 1951. Unfortunately, the UNIVAC-I was delivered too late to be used for the 1950 census, although it found fame for its role in analyzing election outcomes in the 1952 national elections.

Unlike ENIAC, the UNIVAC-I processed digits serially, and this produced a great improvement in its operating speed. The new machine had the capability of adding or subtracting two ten-digit numbers at the then impressive rate of 100,000 operations per second. In addition, the UNIVAC-I CPU operated with a clock speed of 2.25 MHz, which was a significant achievement for a computing machine constructed from vacuum tube circuits.

The UNIVAC-I design put a strong priority on input/output capabilities since its census applications would stress input and output of data over internal data processing. As part of the system, a magnetic tape digital recording unit was developed to provide high speed data delivery to the UNIVAC-I with data rates reaching 40,000 binary digits (bits) per second. At its introduction, the UNIVAC class of computers represented the new state of the art as it captured a majority of the market for general purpose electronic computer systems. As the first commercially available electronic computing system, there was essentially no competition.

In 1955, Remington-Rand and Sperry Corporation merged to become the new Sperry-Rand Corporation. In this transformation, the Univac Division maintained its separate identity, becoming the Univac Division of Sperry-Rand. Eckert remained with Sperry-Rand as an executive and continued with the company as it later changed its name back to Sperry, then merged with the Burroughs Corporation to become Unisys. Having retired in 1989, J. Presper Eckert died in 1995.

Mauchly remained a player in the field of computer science for the remainder of his life. He participated in the founding of the Association for Computing Machinery (ACM) and served as the president of this professional organization. He was also involved in the founding of the Society of Industrial and Applied Mathematics (SIAM). Until 1959, he served as the director of Univac Applications Research, when he left to form Mauchly Associates, a computer applications consulting company. Then, in 1967, he founded Dynatrend, another organization centered on computer consultation. For some years, he also served as a consultant to Sperry. John Mauchly died in 1980.

With the success of the UNIVAC class of computer systems, and seeing great potential opportunities in digital computation, many other commercial firms soon entered into the business of digital electronic computers. However, the limitations of vacuum tube technology created severe constraints on the reliability and practicality of these systems. Although relatively cheap and simple in operation, vacuum tubes, like light bulbs, consume large amounts of energy, generate lots of heat, and have a tendency to burn out, frequently at inopportune times. In addition, they are bulky and slow. All the trends leading toward future computer systems were in the direction of faster and more compact components. It was evident that something better than the vacuum tube was needed.

This growing need for something smaller that wouldn't generate so much heat and would allow for scale-up to permit greater complexity would fuel the next step toward modern computer and electronic technology. The next major step would be the 1947 introduction of the transistor. The transistor would quickly replace the inefficient vacuum tube with a much smaller and more reliable component.

THE TRANSISTOR

Even if it was not recognized as such at the time, when Lee de Forest developed the triode vacuum tube, the door was opened for the world to begin to enjoy the benefits of widespread use of electronic technology. In what is perhaps the next most important development, in 1947, three men at Bell Telephone Laboratories, John Bardeen, Walter Brattain, and William Shockley, developed the transistor, a solid state device that could mimic the characteristics of de Forest's vacuum tube. The three subsequently received the Nobel Prize for their momentous and far-reaching discovery. The original patent name for the invention of the transistor was *three-Electrode Circuit Element Utilizing Semiconductive Materials*.[6]

Among these three men, widely different talents were represented. Brattain was a tinkerer who could make a contraption obediently perform; Bardeen was a theorist who could explain the incomprehensible; and Shockley was a visionary who could anticipate the need for new technology such as the transistor. All three were outstanding scientists, and their unique skills brought them together at Bell Labs. They shared a few years of brilliance, fame, and fortune, and then a clashing of egos sent them on their separate ways.

[6]J. Bardeen et al., U.S. Patent 2,524,035, "Three-Electrode Circuit Element Utilizing Semiconductive Materials." Application filed in 1948, patent issued to Bell Labs in 1950.

The paths of the three men crossed at Bell Labs shortly after World War II, when Shockley had been given responsibility to develop a solid state amplifier while Brattain and Bardeen were assigned as his team members. Brattain served as the experimentalist while Bardeen interpreted results and matched them to theory. Shockley provided direction to the team and monitored the work. Initially, it was an ideal arrangement. But when they obtained their goal of a working transistor, the apportionment of credit became a contentious issue.

The big breakthrough in solid state computing technology came just two days before Christmas in 1947, when the three scientists successfully tested a new germanium crystal device that was intended to act as an amplifier—a device that could replace the vacuum tube in its ability to amplify an applied voltage. The device they tested was one of a class of solid state devices that was designated by the term *transistor*, short for *transfer varistor*. These tiny crystals, made of semiconductor material, were found to have the ability to act like switches, controlling the flow of electricity.

Semiconductor devices would quickly replace the vacuum tube. They were much smaller and more reliable than vacuum tubes and could readily be scaled up in complexity. They didn't give off as much heat as tubes did, so they could be packaged close together. They had no moving parts, so they were less likely to fail. And perhaps most important of all, semiconductors were cheap to make. The first ones were made out of crystals of germanium. Later, silicon became more popular.

Soon after this discovery, Shockley left Bell Labs to return home to Palo Alto in the Santa Clara Valley of California. There he formed his own company, Shockley Semiconductor, in the heart of what would later be known as the Silicon Valley. Bardeen and Brattain also left Bell Labs for positions in academia. Other companies soon began to recruit and hire other Bell Labs star scientists and began turning out semiconductors, including Texas Instruments. Shockley's company failed, however, to ever produce a viable commercial product. It has been said that this was primarily due to Shockley's poor leadership.

Like so much in modern life, some people have missed the true nature of genius and creativity. It is common to think of genius as a quality one is born with, a quality that will follow a person through life. It is worthwhile to consider whether or not that is an accurate picture.

It is interesting to note that John Bardeen was the first to receive two Nobel Prizes in the same field. He was awarded his first one along with collaborators William Shockley and Walter Brattain for their work on the transistor, although it was Shockley who received the most attention for this accomplishment.

Bardeen received his second Nobel Prize for his role in developing the theory of superconductivity, an accomplishment that had been unsuccessfully sought by many other leading theorists, including Albert Einstein, Neils Bohr, Werner Heisenberg, and Richard Feynman. Although Bardeen's work has impacted the world more profoundly than has that of many other well-known scientific geniuses of our time, few people have heard of John Bardeen.

Bardeen did not match the stereotype for a genius. He was modest in speech and demeanor, an ordinary person who happened to be good in science and mathematics. He had the hard-work ethic of a Midwesterner, but he also valued a well-balanced life; he enjoyed such activities as having a picnic with his family. In his collaboration with his partners, he was quiet and cooperative. He does not fit the conventional image of a quirky genius.

Regardless of the personalities of the inventors, it was the ability of transistors to effectively serve as ON–OFF circuit elements, that could be connected in ways analogous to vacuum tubes while eliminating many of the noted disadvantages of vacuum tubes, that established the new semiconductor innovations as the key to the future evolution of computers.

HOW TRANSISTORS WORK

Although most nonmetallic materials are insulators, being highly resistant to electrical current flow, certain materials, such as silicon or germanium, when properly prepared, can exhibit unusual electrical behavior. The preparation of such materials includes the introduction of impurities into their crystalline structure, impurities such as boron or phosphorus, and sandwiching different forms of these materials between conductive plates.

Transistors can be seen in terms of their three basic components, the base, the collector, and the emitter. In reality, transistors are generally tiny pieces of semiconductor material that are fabricated in such a way as to create these three components as separate zones or regions within the block of semiconductor material, using this sandwiching process.

In such a configuration, the base acts as the controller for the electrical output of the transistor. The collector acts as the electrical supply for the device, while the emitter provides the outlet for the electrical flow supplied by the collector. In actual operation, by applying varying (small) voltages to the base, the flow of electrical current from the collector to the emitter can be controlled. As a result, a small change in the current or voltage at the base can result in a large change in the current passing through the device as a whole. Because of this unique electrical behavior, semiconductor transistors can be made to act as either amplifiers or switches. These are precisely the same desirable electronic characteristics of vacuum tubes that helped bring about the era of the digital computer.

Thus the transistor offered a technological jump by performing the same functions as the vacuum tube while eliminating many of their undesirable features—unreliability, bulk, and high energy consumption. And by allowing for significant miniaturization, transistors serving as switches could be interconnected and assembled into ever more complex electronic systems—thus enabling dramatic improvements in the ability to connect ON–OFF circuits.

PROOF OF PRINCIPLE FOR THE TRANSISTOR

The development of the transistor followed the second type of pattern of discovery. Here Shockley, Brattain, and Bardeen used the known phenomenon of the electronic amplification of vacuum tubes as a guide to invent a proof of principle application for semiconductor material.

Shockley believed that he deserved the exclusive and complete credit for the patent because he had directed the initial idea for the transistor. When he explained his position to Bardeen and Brattain a rift developed. Consequently, when the initial patent application for the transistor was submitted, only Bardeen and Brattain were identified as the inventors; Shockley was not included.

However, a short while later, Shockley came up with a significant improvement on the basic idea; he introduced the concept of the junction transistor, a variation that would prove to have much greater commercial value than the original concept, known as the point-contact transistor. While developing this new concept, Shockley excluded Bardeen and Brattain from the effort. Thus the emotions generated by the dispute over ownership of the original idea destroyed the possibility that the three men could work together.

Computers based on solid state transistor technology are generally considered to be Second Generation computing systems. Such computers dominated the commercial computer business from the late 1950s until the early 1960s. Even though the use of transistors and printed circuit technology was a great improvement over earlier approaches, these systems were still quite bulky and energy inefficient. Most such systems were large in size and operated by central organizations such as universities and government agencies.

Thus the world was ready for another leap in technology, and this came in the late 1950s, when it was found that networks of transistors could be etched onto a single piece of silicon with thin metallic connectors. These networks, called integrated circuits (ICs), or chips, became the foundation for the next period of rapid growth in modern electronics.

Computers, meanwhile, got smaller, faster, and more powerful. IBM dominated in the 1950s. The other makers of large, mainframe computers, much smaller than IBM, were known as the Seven Dwarfs—RCA, General Electric, Honeywell, Burroughs, NCR, Sperry Univac, and Control Data Corporation. Even though computers were getting smaller at a rapid pace, by today's standards, they were still quite large and, more importantly, expensive. More and more, scientists and engineers wanted computer systems they could operate themselves, computers that were smaller, cheaper, and easier to maintain.

The introduction of such electronic technologies had already begun and would soon be transforming the world. New innovations led to the packaging of solid state electric components and circuits in ever smaller packages: ICs and chips. The progress in miniaturization of electronic circuits made possible a computer that was smaller, cheaper, and easier to maintain—but another key need was the further development of IC or chip technology and ultimately the development of the "computer on a chip" yet to come.

THE MICROPROCESSOR

The development of the transistor provided a great opportunity for miniaturizing computer components as it enabled the development of the IC. It is clear that the world was ready for this advance, necessity being the mother of invention. An indicator that this is true is seen in the fact that two independent and simultaneous efforts resulted in the introduction of similar IC concepts at the same time.

In 1957, engineer Robert Noyce cofounded the Fairchild Semiconductor Corporation, while Jack Kilby, also an engineer, started working for Texas Instruments. Between 1958 and 1959, both engineering innovators focused their efforts on the key problem of how to fabricate more electrical components from less material.

The ultimate result was the IC and the associated microprocessor chip. In designing a complex electronic machine like a computer, it was always necessary to increase the number of components in order to add capacity. The new IC technology allowed a dramatic increase in the density of electronic components and the complexity of their interconnection on a surprisingly small platform of semiconductor material. Thus the monolithic IC replaced the complex interconnection of discrete components such as resistors, capacitors, and transistors, as well as the associated wiring, by integrating these circuit elements onto the semiconductor platform. Although Kilby's IC was based on a germanium platform and Noyce's on a silicon platform, their ideas were quite similar. The original IC, built by Kilby at Texas Instruments, connected just one transistor, three resistors, and one capacitor.

We can recognize this as the third type of discovery pattern. Here, Noyce and Kilby used a known phenomenon and the already successful proof of principle of the transistor as inspiration for the IC, which combined many elements into a single application.

By 1959, both inventors had applied for patents. Kilby and Texas Instruments patented the technology for miniaturized electronic circuits, while Noyce and the Fairchild Semiconductor Corporation patented the silicon-based IC. In an unusual case of corporate cooperation (but only after several years of litigation), the two companies agreed to cross license their technologies, opening the way for creation of a global market.

Kilby subsequently received the Nobel Prize in the year 2000 for his part in the invention of the IC. In addition, he holds patents on more than sixty inventions. He died in 2005.

Noyce went on to cofound Intel Corporation, the company that developed the microprocessor in 1968. But for both men, the invention of the IC was their proudest accomplishment. Most modern products use IC technology, and this technology must be considered to be one of the most influential innovations of the 20th century.

Shortly after the first laboratory demonstration of IC technology, Fairchild Semiconductor Corporation produced the first commercially available ICs. Almost immediately, new computer systems began using IC technology rather than discrete components.

Third Generation computers using new IC technology experienced explosive expansion in computer use beginning around 1963. Computer development proceeded in two very different directions. The manufacturers of large mainframe computers, such as IBM's 360, capitalized on the new technology to dramatically increase the process and storage capacity of their state-of-the-art machines; at the same time, the introduction of IC technology spurred the introduction of the minicomputer, a much smaller, decentralized, multiuser computer that allowed many smaller businesses to benefit from advanced computing resources.

In addition, large-scale circuit integration allowed for the development of many new specialized applications such as special purpose processors intended for analyzing flight data in military fighter jets. These increasingly small dedicated processor units were only a taste of things to come.

Early ICs consisted of small numbers of transistors and other circuit elements; however, by 1970, the circuit density of ICs had grown to thousands of elements per wafer. Eventually, the time would come to fabricate an entire computer on a chip. In fact, in 1970, the Intel Corporation introduced the 4004 chip, a 4-bit microprocessor; and within two years, they introduced the 8008, an 8-bit microprocessor.

And so this brings us to the Fourth Generation computers, computers that use microprocessor technology by incorporating much of the computer's processing abilities on a single chip. When combined with the rapidly improving memory storage chip, the RAM chip, these microprocessor-based computers were becoming smaller and faster than ever.

And since the 1970s, the processing power of microprocessors has relentlessly improved. However, the underlying technologies of LSI (large-scale integration) and VLSI (very large-scale integration) has remained fairly constant.

So where is the relentless improvement to be found? While the physical size of chips had actually grown between 1960 and 1975 (some of the more complex chips were up to a factor of 20 larger in 1975 than 15 years earlier), the density and complexity of the circuits contained on the chips had increased even faster, by approximately a factor of 32 over the same period. Considering these changes in both directions, an overall change of a 640-fold improvement in the number of components (as predicted by Moore's Law) is accounted for by both reduced circuit element size and improved packing efficiency of the components.

HOW MICROPROCESSORS WORK

In computer design, it is common to refer to the *logic gate* as the basic element. Although the term *gate* creates a connotation of a doorway through a fence or wall that may swing open and closed, nothing like this mechanical motion occurs on a circuit chip. Instead, the function of a gate is to control the flow of electricity in a circuit, either by increasing the flow to amplify a signal, or by acting as an ON–OFF switch to represent logic steps or data. This functionality is similar to that of the electric flow control in a vacuum tube.

In a computer, logic gates are used to control flow of electricity in a circuit. The logic gates are created by the transistor as the basic unit. Transistors of two types are used by computer designers: these are PMOS (*p*ositive-channel *m*etal-*o*xide *s*emiconductor) and NMOS (*n*egative-channel *m*etal-*o*xide *s*emiconductor) transistors. These two elements differ in that the NMOS transistor is turned on or off by the flow of electrons, while the PMOS depends on the flow of electron vacancies, similar to a flow of positive charge.

Current flowing through a given gate determines the state of that particular gate; the voltage of the gate represents a single bit of information. A high voltage represents the value "1" while a low voltage represents a "0." Ultimately, the control of current flow in the logic gates amounts to control of the flow of information in the complex circuit.

The computing power of logic gates comes from the output of any particular gate as a voltage that can be used to control another gate. In other words, the functionality of the computer as a complex collection of circuit elements comes from the interconnection of those elements and the switching function that the logic gates provide.

The evolution of inventions from vacuum tubes to transistors to microprocessors has created a revolution in computing. One of the primary measures of the strength and impact of this revolution can be seen in the Moore's Law observation of the exponentially increasing number of components contained on chips.

MOORE'S LAW

More than forty years ago, Gordon Moore, the coinventor of the IC, published an article in which he considered the future development of the semiconductor industry. He reported his observation that the complexity of the lowest cost semiconductor components had essentially doubled each year since 1959 when the first microchip was introduced. This observation of an exponential rate of increase in the number of components on a chip subsequently became known as Moore's Law. In a later refinement, Moore's Law has become accepted as the prediction that a chip's capacity would double every 18 months. See Figure 2-1, which shows the year and basic component technology on one axis and

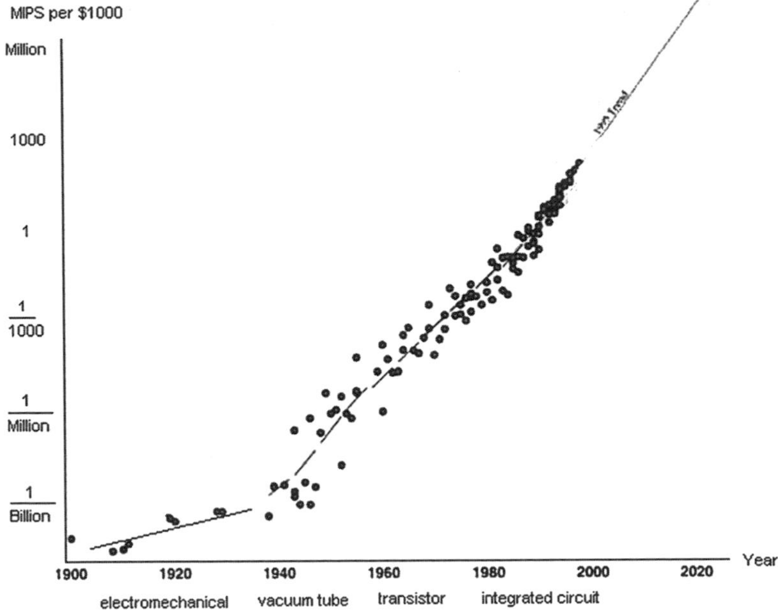

Figure 2-1 *Evolution of computer power/cost.*[7]

the cost/performance ratio (in units of millions of instructions per second (MIPS) per $1000 cost) on the other axis.

As impressive as its predictive power has been, the exponential growth predicted by Moore's Law and indicated by extrapolation of the lines on Figure 2-1 may ultimately prove to be unsustainable as practical and theoretical limits are approached. Thus the growth of Moore's Law, as with other projections of technological advancement, must be considered in light of a more comprehensive pattern of technology—introduction, growth, and maturation—like the S-shaped curve.

Moore's Law has primarily focused on information processing technology; however, the impacts of the underlying technology changes reflected in the Moore's Law observation go far beyond computers and data processing applications, creating a broad range of economic, organizational, and social impacts. In reality, many planners sometimes disregard the future cost of computing resources in their long range planning.

Regular doubling produces exponential growth. Exponential growth, however, also means that the fundamental physical limits of microelectronics are being approached ever more rapidly. Several observers have therefore speculated about the possibility of the end of Moore's Law.[8]

[7]This graph is based on a similar presentation by H. Moravec, "When Will Computer Hardware Match the Human Brain?" *Journal of Transhumanism*, Vol. 1, March 1998.

[8]See I. Tuomi, "The Lives and Deaths of Moore's Law," *First Monday* 7(11): November 2002. `http://www.firstmonday.org/issues/issue7_11tuomi/`.

While many have speculated over the future of Moore's Law, some have concluded that it will probably be valid for at least a few more generations of technology, or about a decade. The International Technology Roadmap for Semiconductors (ITRS),[9] a planning effort that considers developments through 2016, can be considered to represent a consensus view of the potential for future technology advances.[10]

ITRS suggests that, within the next 10–15 years, "most of the known technological capabilities will approach or have reached their limits." Nevertheless, its basic premise is that Moore's Law, although it may slow down, will still provide an appropriate basis for projecting future developments in the electronics industry.[11]

During its history, the semiconductor industry has several times appeared to reach a limit. However, in each case a new factor has emerged to continue the basis for technological expansion and fulfill the projections of Moore's Law. Some of these new factors include the digital clock and calculator industry; the mini and mainframe computer industry; the personal computer; and the impacts of the Internet and the World Wide Web.

PATTERNS OF DISCOVERY

In the story of Moore's Law, we followed the tributary inventions—the vacuum tube, transistor, and microchip—feeding the development of the computer and extending Moore's Law. Edison, Fleming, de Forest, Mauchly, Eckert, Bardeen, Brattain, Shockley, Noyce, Kilby, and Moore all helped connect circuits and opened the floodgates to information processing.

The phenomenon of producing light from electrically generated heat was recognized and the proof of principle experiments establishing the potential of electric lighting systems were performed by 1871. Edison was inspired by his insight that a high resistance filament, in a parallel circuit using newly developed high vacuum technology, could produce an incandescent light. He persevered until he achieved it. Along the way he serendipitously discovered the Edison Effect—a phenomenon of controlled electron flow.

Fleming used the Edison Effect to invent a proof of principle application—the vacuum tube diode. Building on these discoveries, de Forest was inspired to modify the controlling elements of the vacuum tube to produce the electronic amplifier.

Mauchly and Eckert's inspired design of ENIAC connected the circuits of many vacuum tubes.

Bardeen, Brattain, and Shockley used the known phenomenon of the electronic amplification of vacuum tubes as a guide to invent a proof of principle application for semiconductor material.

[9]"International Technology Roadmap for Semiconductors (ITRS)." `http://www.itrs.net/about.html`.
[10]See I. Tuomi, "The Lives and Deaths of Moore's Law," *First Monday* 7(11): November 2002.
[11]Ibid.

The problem each inventor solved was how to control electronic circuits to produce ON and OFF signals within logic circuits efficiently.

Miniaturization and mass production improved the efficiency of connecting circuits according to Moore's Law. The microprocessor combined logic circuits in a small, efficient, and inexpensive unit for processing logic and information. The fabrication process now permits a steady stream of development.

The vacuum tube and the transistor were valuable as switches from which logic circuits could be formed. But it was miniaturization of these switches by many orders of magnitude that created the microchip and opened the floodgates of the Information Revolution. We will look to this success when we ask why software has failed to produce similar orders of magnitude improvements in its ability to process data.

Each of these inventions was the result of a particular pattern of discovery. Edison's serendipitous discovery of the Edison Effect was followed by three generations of computer technology based successively upon the vacuum tube, the transistor, and the microprocessor—each following the Proof of Principle Pattern.

But patterns of discovery should not be only explored as isolated events. They can also provide insight when considered as collections forming a pattern of patterns. The inventions of the vacuum tube, transistor, and microprocessor illustrate cascading generations of computing enabling a continuation of Moore's Law through several sequential technology development periods.

FORECAST FOR CONNECTING CIRCUITS

Starting from our analogy at the beginning of this chapter, recall that each spring the snow packs of the Appalachian and Rocky Mountains melt, providing growing flow to the tributary streams of the Mississippi. As the flow provides navigability and transportation of goods to New Orleans, it also improves the economy of the United States.

In a similar cycle, we can consider Moore's Law in which every 18 months of innovation refreshes the chip industry, speeding the flow of data and fueling the Information Revolution. Most of the innovations are inspired modifications of proven principles. However, occasionally new principles are discovered.

In order to forecast the future development of the microprocessor, let's explore prospects for continued growth following Moore's Law (see Figure 2-1). We need to consider three possible alternatives:

1. Inspired modifications of Moore's Law constrained by physical limits.
2. New principles overcoming limitations and extending Moore's Law.
3. Serendipitous discoveries that create new opportunity for Moore's Law to continue unabated into the distant future.

Discoveries Requiring Inspiration and Perspiration

One of the valid concerns related to future technology development is the possibility that Moore's Law may be dissipated in the near future. The technical and physical obstacles that loom ahead are of concern. How complex can chips become is an open question. One problem is power leakage as chips become more power hungry. Chips with billions of transistor can leak up to 70 watts of power, causing cooling problems.

In addition, it is possible that circuit dimensions cannot get much smaller than the current 65 nanometers without increasing production difficulties.

In 2004, we were mass producing the 90-nm IC. By 2006, migration to 65 nm began. Changing from 90-nm to 65-nm design rules was quick because it required no changes in the fabrication process.

Back in 1988, IBM fabricated the world's smallest transistor at that time using 70-nm design rules. It ran on a power supply of only 1 volt rather than the usual 5 volts, but required nitrogen cooling. Today these field effect transistors (FETs) run at room temperature.

In late 2003, NEC built a FET with a 5-nm gate length and IBM built one with a 6-nm gate length. These are 10 times smaller than what's used in production now. So we already know that transistors will work at scales that are an order of magnitude smaller than today's technology.

Instead of focusing on acquiring more speed from a single processor, which could be a dead end, innovators are also developing multicore processors: microprocessors using four to eight parallel "cores." By working in parallel, the total throughput of the processor is greatly increased. Dual and quad cores are already being produced commercially. Parallel advances in hardware will optimize the threaded software of network services.

Discoveries Requiring New Proof of Principle

What about extending Moore's Law? If we reach a physical limit to Moore's Law, we will require new discoveries in breakthrough phenomena demonstrating new proofs of principle. The following are some of the leading candidates.

3D Molecular Chips One alternative is building 3D chips, in which layers of transistors form a high rise.[12] New technologies that could lead to molecular three-dimensional computing include nanotubes, nanotube molecular computing, and self-assembly in nanotube circuits.

Matrix Semiconductor, Inc. is already building three-dimensional circuits using conventional silicon lithography.[13] They are manufacturing memory chips with vertically stacked planes.

[12] M. Kanellos, "Intel Scientists Find Wall for Moore's Law," *CNET News.com*, published by ZDNet News, December 1, 2003.

[13] T. H. Lee, "A Vertical Leap for Microchips," *Scientific American.com*, January 2002.

Carbon nanotubes and silicon nanowires can be metals or semiconductors that are extremely strong materials with good thermal conductivity.[14] These characteristics can be used as nanowires or field effect transistors. Carbon nanotubes can be 1–2 nm in length, and their use could result in substantial reductions in energy consumption.

In 1991 the first nanotubes were assembled by rolling a hexagonal network of carbon atoms into a cylinder. In a demonstration at the University of California at Irvine by Peter Burke, nanotube circuits at 2.5 gigahertz (GHz) were operated. Such circuits may have a theoretical speed that could be 1000 times faster than today's computers. Since this effort, significant additional achievements have been accomplished, and operation of nanotube-based circuitry has been demonstrated at frequencies of up to 10 GHz.[15]

In The Netherlands, Adrian Bachtold and others at Delft University of Technology are also working on nanotube transistors.[16] Also, researchers at the University of California at Berkeley and Stanford University have created an integrated memory circuit based on nanotubes.[17] This technology has some problems, such as the fact that nanotubes may be conductive, and there are issues about lining up nanotubes. IBM scientists have already shown that nanotube transistors could be grown in a manner similar to silicon transistors. Thomas Rueckes of Woburn, Massachusetts demonstrated a single-chip wafer with ten billion nanotube junctions properly aligned in 2003.

James R. Heath, Pat Collier, and Eric Wong reported molecular-based logic gates for the first time in 1999. They achieved results at least as good as silicon and with components that would still be fault tolerant. Heath believes there could be a hybrid computer with molecular electronics in a decade.[18]

Defense Advanced Research Projects Agency, the National Science Foundation, and the Office of Naval Research funded the group of chemists who use a class of molecules called rotaxanes—synthetic, dumbbell-shaped compounds for logic operations that can provide memory and routing signals. A critical step in making a molecular computer requires that the wire be arranged in one direction as molecular switches and that a second set of wires is aligned opposite. A single layer of molecules—the rotaxanes—is at the junction of these wires. The chemists showed they could link molecular switches and wires together into a logic circuit.[19]

[14]"Synthesis of Nanowire Heterojunctions for Advanced Nanoelectronic Devices." `http://www.urop.uci.edu/SURP/sample_proposals/SURP%20Physical%20Sciences%202.pdf`.

[15]"Today@UCI," March 9, 2007. `http://today.uci.edu/Features/profile_detail.asp?key=193`.

[16]A. Bachtold et al., "Logie Circuits with Carbon Nanotube Transistors," *Science* 294(5545):1317–1320, November 9, 2001.

[17]S. Yang, "Researchers Create First Ever Integrated Silicon Circuit with Nanotube Transistors," U.C. Berkeley News Release, January 5, 2004.

[18]"UCLA Chemists, Hewlett-Packard Labs Colleagues Report Significant Advances Toward Chemical Computers," *Science Daily*, July 19, 1999.

[19]S. Wolpert, "Dawn of Molecular Computer," UCLA Today. `http://www.today.ucla.edu/1999/990727dawn.html`.

An alternative approach by Fujio Masuoka, the inventor of flash memory, has a memory design that reduces the size and cost per bit by a factor of 10. Rensselaer Polytechnic Institute and MIT have also demonstrated three-dimensional silicon chips working prototypes.

Single-Electron Transistors Major progress has been made in computing using just a few molecules. Avi Aviram of IBM and Mark A. Ratner of Northwestern University first suggested it in 1970. Since then, Christoph Wasshuber of Texas Instruments has developed a single-electron transistor.[20]

Using a single electron to turn a transistor on and off would miniaturize as well as reduce power. There have been severe problems due to their extreme sensitivity to background noise. However, Wasshuber claims to have designed a single-electron transistor that can be incorporated into silicon circuitry but remain immune to interference.

Single-electron transistors could store as much as a terabit of data in a square centimeter of silicon. That is a two order of magnitude improvement over today's technology.

Crossbar Latch The Quantum Science Research (QSR) group of Hewlett Packard has demonstrated a crossbar latch.[21] This technology doesn't use transistors to provide the signal restoration and inversion required for general computing. It allows development of nanometer devices and could improve computing by three orders of magnitude.

The experimental latch is a single wire that lies between two control lines at a molecular-scale junction. Voltage to the control lines allows the latch to perform NOT, AND, and OR operations.

Optical Computing Another innovative concept that could allow for the continuation of Moore's Law advances is the idea of using photons of light instead of electrons. Fiber optics is the better technology for most digital information and communications applications. However, optical circuits face major obstacles to compete with high density microprocessor technology.

Discoveries Requiring Serendipity

It would be appropriate for us to refer to S. Harris's cartoon ". . . and then a miracle occurs . . ." (see page 7) at this point because by its very nature, serendipity is unpredictable. Nevertheless, such discoveries do occur and can set off a whole chain of events just as the Edison Effect did. One area that would require a giant leap beyond merely a new proof of principle to succeed would be quantum computing.

[20]"Nanotechnology," *Encyclopædia Britannica Online*, March 9, 2007. <`http://www.britannica.com/eb/article-236446`>.

[21]M. Singer, "HP's 'Crossbar Latch' to Replace Transistors?" *Internet News*, February 2, 2005. `http://www.internetnews.com/ent-news/article.php/3467491`.

Quantum Computing An exciting and interesting future technology with the potential to dramatically extend the life of Moore's Law (or to replace it with a new paradigm) is the idea of the quantum computer. A quantum computer is a computational device that uses the phenomena of quantum mechanics—the strange and quirky science of the ultrasmall—as a basis for operation. Phenomena such as superposition and entanglement of states would be used to perform operations on data.[22] While conventional computers process information in units of bits (short for binary digits), a quantum computer is based on quantum bit, or qubits, which can take on values of one, zero, or some superposition of these. The basic concept of quantum computing holds that the quantum properties of particles can be used to represent and structure data, opening the possibility for new and unique approaches to the solution of certain classes of problems.

Top business minds, such as Boston-based BBN Technologies, are taking an interest in quantum computing. They join researchers in places such as Harvard, Stanford, the University of Waterloo, the University of Calgary, and DARPA.

[22]"First Quantum Cryptographic Data Network Demonstrated," *Science Daily*, August 29, 2006.

3

Connecting Chips

The best way to predict the future is to invent it.
—Alan Kay[1]

Just about everyone is aware of the impact the personal computer has had on society, the creative result of *connecting chips* in new and productive ways. But do you know who connected the chips that created the first personal computer?

Was it Micro Instrumentation and Telemetry Systems (MITS) who introduced the Altair 8800 in the mid-1970s? In a provocative issue of *Popular Electronics* in January 1975,[2] a picture of the MITS Altair 8800 appeared on the cover, and the article within erroneously trumpeted the 8800 as the first "personal" computer. As a result of the splash this article made, thousands of orders were placed for the 8800, and MITS was able to avoid bankruptcy. Interestingly, this also stimulated Paul Allen and Bill Gates to develop BASIC for the Altair 8800, a boost that resulted in the creation of the Microsoft Corporation.

Was it Apple with the Apple I? Early in 1976, two college dropouts, Steve Jobs and Stephen Wozniak, founded the Apple Computer Company and began operating out of a garage, building the Apple I, which some claim to be the first personal computer to be sold as a fully assembled package. A third partner in Apple, unfortunately, sold his 10% stake in the venture for $800 almost

[1]T. Brandow, "The Future of Computing Is Invention: An Interview with Alan Kay," Hewlett Packard Feature Story. `http://www.hp.com/hpinfo/newsroom/feature_stories/2002/alankay02.html`.

[2]"World's First Minicomputer Kit to Rival Commercial Models," *Popular Electronics*, January 1975.

Connections: Patterns of Discovery By H. Peter Alesso and Craig F. Smith
Copyright © 2008 John Wiley & Sons, Inc.

immediately due to his lack of confidence in the chances that Apple would be a success.

Or was it Xerox with the Alto? Originally conceived in 1972 and operated in 1973, the Alto was intended to be the first general purpose computer for individual use. At the time, most computers were large, multiuser, batch processing machines. A programmer would create a program and assemble it together with input data using punched cards, schedule a time to use the computer, submit the card deck for processing, and return later for results. The Alto concept, developed at the Xerox Palo Alto Research Center (PARC), was the first real-time machine that could be dedicated to a single user so as to be interactive and available for use at any moment.

In developing the Alto concept, the scientists at Xerox PARC created much more than a personal computer. They designed, built, and demonstrated a complete system of hardware and software that fundamentally changed computing itself. An impressive list of "firsts" were produced at PARC; their brilliant engineers developed the first graphics-oriented monitor, the first handheld "mouse" interactive input device, the first word processing program, the first local area communications network, the first object-oriented programming language, and the first laser printer.

Their concept of personal computing challenged the accepted wisdom of how people and computers could interact. Most computer professionals scoffed at the idea of one computer for each person. By the mid-1970s, PARC had crafted a framework of machines and programs that were controlled by individuals and linked through networks, enabling the sharing of resources.

Xerox, however, did not convert its vision of personal distributed computing into either commercial success or the recognition now enjoyed by Apple and IBM. It's not that Xerox failed to profit financially from its innovative technologies; for example, the company's laser printer business thrived and proved highly profitable. But Xerox management perceived the Alto venture to be a journey into the unknown, and they failed to seize the opportunity to define and dominate the world of personal computing.

Even though chips have found their way into nearly every electric device we use, nowhere have they had a greater impact than in the introduction of the PC. The idea of connecting chips into low-cost, multipurpose, personal computers has proved an important transformational idea.

The story of how this transformation came about is as much a story of marketing and corporate decisions as it is of technology innovation and invention. The start-up of PARC in the early 1970s and its record of developing innovative technology solutions set the stage by producing the key inventions leading to the dramatic success of the PC; however, the success of Apple and IBM in commercializing this technology was equally important. The *proof of principle* innovations at PARC led to the *inspiration/perspiration* development of those that followed at IBM and Apple. These events and the innovative individuals whose efforts led to the development of the personal computer are the focus of this chapter.

THE PERSONAL COMPUTER STORY

Three millennia ago, where three continents meet, the most fertile land in the ancient world was called the Fertile Crescent. The Fertile Crescent was the source of some of the oldest civilizations in history, including the Assyrians, the Babylonians, the Sumerians, the Hittites, and the Egyptians.

Silicon Valley is sometimes referred to as the 21st century's fertile crescent, and one research laboratory within Silicon Valley was once known as the most productive developer of technology since Edison's Menlo Park. That laboratory was the Palo Alto Research Center, or PARC, founded by Xerox Corporation in 1970 with the purpose of creating technology options for the future growth of Xerox.

From the time of its invention in the late 1940s through the end of the 1970s, computer technology had been unaffordable, inaccessible, and useless to most people. Computers were very large investments that were owned and operated by large corporations, government agencies, and universities, but not by individuals or small businesses. The large organizations operated a technology that required highly specialized knowledge and provided results for a narrow set of applications. For the most part, computers manipulated numbers for scientists, engineers, and accountants; and they were operated as a centralized support service.

All of this had changed by the 1980s when the successful introduction of the IBM PC and the Apple II and Macintosh proved that there was a robust market and future for the personal computer. In fact, the advertising campaigns of the time emphasized the differing but successful approaches being taken by these two competing corporations that were offering new devices to quench the growing public thirst for personal computing technology.

IBM emphasized consumer education in its marketing strategy, and it rolled out an advertisement with the Charlie Chaplin character "the Little Tramp" as the face of IBM's technology. The idea was that if the Charlie Chaplin tramp character could operate an IBM PC, then the technology must be accessible. The campaign was a remarkable success, and by 1987 Americans had purchased more than 25 million PCs.

In contrast, the upstart company Apple Computer responded to the dominant position of IBM as a corporate giant by running a memorable, though brash commercial for the Macintosh computer. The Apple commercial, which was run officially only once during the 1984 Super Bowl broadcast, showed a dark and sinister world of corporate uniformity, an Orwellian vision, into which an individual female athlete entered and destroyed the television image of the controlling authority. The commercial was, in essence, a video morality play celebrating the glory of iconoclastic individualism while condemning the ominous threat of large organizations (e.g., IBM?) that were capable of oppressing the human spirit. Using imagery without words, Apple drew the battle line clearly between itself and IBM.

But in 1973, long before these corporate titans began their struggle, researchers at Xerox's PARC flipped the switch on the Alto, the first computer designed and built for use by a single person. Xerox scientists as well as secretaries were using personal computers that were superior to any system sold in the market long before 1984, the year of the Apple Super Bowl commercial.

The Alto was the product of hard work, creative innovation, and a vision of the future of technology that saw the value of placing computing resources directly in the hands of individual users. Who were the key people most responsible for the groundbreaking Alto initiative, and where did this vision come from? Some of the key players included Vannevar Bush, Douglas Engelbart, Robert Taylor, J. C. R. Licklider, Alan Kay, Butler Lampson, and Charles (Chuck) Thacker.

So the Alto actually starts with the thoughts and ideas of Vannevar Bush.

VANNEVAR BUSH

In 1945, Vannevar Bush, a prominent advisor to President Roosevelt, published a landmark article entitled "As We May Think,"[3] in the *Atlantic Monthly* that described a concept he had been contemplating since at least the early 1930s. This concept, which he called "Memex," was a microfilm-based automatic information handling machine that would allow an individual to store, organize, and process all types of documents, records, and communications. It would be a machine to augment the user's memory and establish associative links between records that would enable meaningful sorting and processing of information.

Bush's Memex machine would consist of a desk, viewing screens, a keyboard, input devices such as levers and buttons, and a data storage device based on microfilm technology. As an information processing system dedicated to a single user, the parallels with the PC are clear. The Memex machine was far ahead of its time, and Bush's concept is generally credited with being the inspiration behind two major technologies of the current information age: hypertext technology (essential in the development of the World Wide Web) and the personal computer.

Bush's ideas were compelling, and it was Douglas Engelbart who took the next step.

DOUGLAS ENGELBART

Today, Douglas Engelbart is well known for his part in inventing the computer mouse and the graphical user interface (GUI), two key components of personal computer technology, but in 1948, he had just completed his degree in electrical engineering when he experienced an epiphany. He started thinking

[3]V. Bush, "As We May Think," *The Atlantic Monthly*, July 1945.

about ways in which a machine could be built that would augment human intellect in accordance with Vannevar Bush's concept of the Memex.

In contrast to Bush's idea, Engelbart envisioned a personal information processing device that would feature a TV monitor output system where the user could visualize models of information in graphic display and move around with a pointing device to explore the model dynamically.

Engelbart had always been a forward thinker. During the 1960s, he advanced new ideas about the use of computers for work conferencing and collaboration. Following the lead of Bush, he saw the promise of using technology to leverage the individual human intellect. He believed that technology could provide solutions to many of the complex problems facing people in the modern world.

Born in 1925 in Oregon, Douglas Engelbart was raised on a small farm during the period of the Great Depression. After graduating from high school in 1942, he enrolled at Oregon State University, where he majored in electrical engineering. The outbreak of World War II interrupted his academic studies, and he entered military service with the U.S. Navy as a radar technician. During his two-year assignment in the Philippines, he had the opportunity to read Bush's article "As We May Think." This article made a great impression on him and was to become a factor influencing his future technology interests and directions.

Following his wartime service, he returned to the university and completed his undergraduate education, receiving his B.S. degree in electrical engineering in 1948. He then took a position at Ames Laboratory in Sunnyvale, California, working for NACA (the National Advisory Committee for Aeronautics, the predecessor to the current NASA).

On a personal level, Engelbart's attitudes gradually matured from an interest in "a steady job, getting married and living happily ever after," to a drive to contribute to society as expressed in his question: "How can my career maximize my contribution to mankind?"[4]

His study of Bush's Memex concept brought him to envision people using graphic displays that would represent an information space that could be explored and used to "formulate and organize their ideas with incredible speed and flexibility." Realizing that he would need to expand his technology capabilities to accomplish the goal of maximizing his contribution to mankind, he enrolled in graduate school at the University of California at Berkeley to continue his studies in electrical engineering. He received his Ph.D. in 1955, and after a short period as an assistant professor at UC Berkeley, he accepted a position at the Stanford Research Institute (SRI).

In 1962, Engelbart prepared a forward-thinking article entitled "Augmenting Human Intellect: A Conceptual Framework."[5] This paper was the product

[4]S. Griffin, "Internet Pioneers: Douglas Englebart." `http://www.ibiblio.org/pioneers/englebart.html`.

[5]D. Englebart, "Augmenting Human Intellect: A Conceptual Framework," Stanford Research Institute Summary Report AFOSR-3233, October 1962.

of his work in support of the U.S. Air Force Office of Scientific Research. In it, he described his ideas for using computer technology to enhance human capabilities. As an example, he introduced the concept of graphical computer-aided design (CAD) software for the design of buildings and structures. This was a huge leap forward in thinking.

Engelbart's focus on the idea of a graphical output display and the associated ability to explore the "information space" led to the invention of the graphical user interface (GUI). With his GUI concept, an analyst could use graphical elements, instead of text, for the input and output of a computer program. And his related invention of the computer mouse, or electronic pointing device, was an essential adjunct to the GUI since it allowed the analyst to move a pointer around on the screen and select command options rather than requiring him/her to type commands into a text device.

Beginning in 1963, Englebart started his own research laboratory within SRI, the Augmentation Research Center. Much of his work during the 1960s and 1970s was devoted to the implementation of hypertext ideas and the creation of digital libraries of documents that could use hypertext linkages to enhance storage, processing, sorting, and retrieval—clearly ideas influenced by the Memex concept. In addition, during this period, several other innovations were introduced, including the mouse, the GUI, and the capability for on-screen video teleconferencing.

In 1968, Engelbart organized a technology demonstration that would have lasting repercussions. During the Fall Joint Computer Conference in San Francisco, he and his 17 staff researchers presented a 90-minute live multimedia demonstration that displayed the ability of videoconferencing and remote collaboration with individuals at a distant site some 30 miles away. The demonstration was well attended with some 1000 computer professionals witnessing the event. In addition to the demonstration of videoconferencing and remote collaboration, the presentation also showcased several other milestones in technology development, including the mouse, the GUI, and hypertext.

This demonstration opened people's eyes to the technology possibilities that were just emerging. It showed that the world could collaborate on electronic documents displayed on computer screens and transmitted instantly over networks. The demonstration also stimulated a response from Xerox; fearing the impending demise of their paper-based business as the world moved to a paperless future, Xerox management decided to form the Palo Alto Research Center, or PARC, in 1970. With strong financial backing, PARC quickly became one of the top research and development facilities in the country.

Thus not only did Engelbart contribute directly to the technology needed for the development of the personal computer, he also stimulated the formation of PARC, one of the most productive and creative sources of computer science innovation.

When PARC was established, Xerox made sure that it included leadership that was capable, innovative, and visionary. In this role, one of their most important selections was Robert Taylor.

ROBERT TAYLOR

Born in Texas in 1932, Robert Taylor was the son of a Methodist minister. After completing high school, he enrolled at Southern Methodist University. Following military service during the Korean War, he went back to resume his education under the GI Bill. When he finally completed his master's degree in experimental psychology, he decided to stop his formal education since he had no interest in pursuing a Ph.D. in the field of psychology.

Having finished his academic training, Taylor found employment in several engineering positions at various aerospace companies. While pursuing this line of work, he had the opportunity to write a proposal to the National Aeronautics and Space Administration (NASA) for a flight control simulation display system; as a result of this proposal, he soon found himself working for NASA in Washington DC.

While at NASA, he become keenly interested in the rapidly growing field of computer science, and here he also had the opportunity to meet and interact with two other visionaries in the field: Douglas Engelbart, who was engaged in cutting-edge computer science research at SRI; and J. C. R. Licklider, who was the head of the Information Techniques Processing Office (ITPO) at the Advanced Research Projects Agency (ARPA). The connections to these two other individuals would turn out to be important: while at NASA, Taylor was to fund Engelbart's work at SRI in computer display technology leading to the development of the computer mouse, among other innovations; and ultimately Taylor would succeed Licklider in his position at ARPA.

While at NASA, and subsequently ARPA, Taylor earned a reputation for having strong vision in pursuing projects, and extraordinary skill in assembling and leading teams of scientists. Although he did not have the computer science training to conduct research himself, he became very interested in the topic of time sharing and soon gained a sophisticated feel for the future of digital computing technology. Having met and interacted positively with Licklider, Taylor soon took a position at ARPA that led to his assignment as director of programs in computing. Eventually, he would take over as director of the ITPO at ARPA after Licklider departed that agency.

When the decision was made to start up the new Xerox research center PARC, Xerox management realized that the ability to recruit and lead a talented team of scientists was crucial. They turned to Taylor to be part of the PARC management team and enlisted him as manager of the Computer Science Laboratory (CSL) at PARC. Taylor demonstrated his ability to lead with resolute vision and showed his great skill in assembling and leading the teams of talented scientists who made it all happen at PARC. In particular, he was successful in recruiting key staff from NASA, ARPA, SRI, the University of Utah, Carnegie-Mellon, and MIT. His hiring of technology leaders Alan Kay and Butler Lampson was to be critical.

While at PARC in the 1970s, Taylor managed the development of the principal technologies for the personal computer and computer networking. He would become a key player in the development of the ARPAnet, the predecessor of the Internet. In fact, his great interest in the technology of interactive computing led him to state: "Interactive computing is like a conversation. It's a huge change. Now I want to interconnect them. That's what ARPAnet is all about."[6] In 1999, Taylor was awarded the National Medal of Technology "for visionary leadership in the development of modern computing technology, including computer networks, the personal computer and the graphical user interface."[7]

Taylor owed much of his vision to his predecessor and colleague, J. C. R. Licklider.

J. C. R. LICKLIDER

Born in 1915 in Saint Louis, Missouri, J. C. R. Licklider attended Saint Louis University, where he received a triple degree in the fields of mathematics, physics, and psychology. He exhibited particular interest in the operation of the human brain and went on to complete his doctoral studies at the University of Rochester with dissertation research in the area of brain function. This background would serve him well in his future career in advanced research related to computing and networking.

In the 1960s, computer experts were complaining because they could not interact with computers when and how they wanted. Licklider argued that the principal function of the computer should be improving human thought. In 1968, he along with Robert Taylor published a paper entitled "The Computer as a Communication Device"[8] that laid out the future of the Internet. The paper starts out: "In a few years, men will be able to communicate more effectively through a machine than face to face." Licklider and Taylor both realized early on that the role of the computer as a communications device could easily eclipse its role as a numeric processor.

At the time Licklider began his career, computers were large and cumbersome, and they were operated by computer specialists in centralized organizations. The computer users wrote their programs, transferred them onto punched cards, and submitted their card deck to be loaded into the computer as part

[6]M. Softky, "Building the Internet," *The Almanac*, October 11, 2000. `http://www.almanacnews.com/morgue/2000/2000_10_11.taylor.html.`

[7]Ibid.

[8]J. C. R. Licklider and R. W. Taylor, "The Computer as a Communication Device," *Science and Technology*, April 1968.

of a batch of such decks. The results would be returned to the users a day or so later. Any errors in the card deck would have disastrous results. The smallest keypunch error would result in the program not running. It would have to be resubmitted for the next day.

One of the research areas that interested Licklider was the technology of time sharing, where multiple users could share computing resources by working at separate terminals and receiving slices of computer execution time as they were available. With high speed computers, the user would not be aware that he/she was sharing the resource since the response time would be nearly immediate. This innovation also resulted in the ability to use a computer in an interactive way, a major change in the way researchers related to their computing resources.

After leaving ARPA in 1975, Licklider returned to academia, where he served as professor at the MIT Laboratory for Computer Science (LCS) until his retirement in 1986. He passed away in 1990, having left a legacy of great accomplishment and vision that continues to impact the development of computer technology to this day.

ALAN KAY

A pioneering expert on graphics applications, Alan Kay was among the first members of the ARPA research community to hear about Robert Taylor's move to PARC. It was no surprise when this influential computer scientist also left to take a position at PARC.

Kay was born in 1940 in Springfield, Massachusetts and was well traveled as a youth. When he was a small child, he lived with his family in Australia, returning after a few years because of conditions related to the onset of World War II. He was a gifted child, and having learned to read by the age of three, he claims to have read hundreds of books by the time he entered school.

Books and music had dominated his childhood. An avid reader, he also played jazz professionally as a teenager. He scorned traditional education. Although he was one of television's original Quiz Kids, he nearly failed the eighth grade. Following high school, Kay continued his jazz career, flirted with college, and then joined the Air Force, where he learned how to program computers. His interest stimulated, he returned to his education in science after completing his military service, eventually turning to the field of computer science. By 1969, he had completed his university studies, receiving his Ph.D. degree in computer science from the University of Utah. In his doctoral thesis, Kay described an extraordinary programming language and computing machine called FLEX, an early desktop computer with graphical user interface and object-oriented operating system. Following his academic studies, he took a position at the Stanford Artificial Intelligence Laboratory, where he worked on concepts for the notebook computer as well as the design of the Smalltalk language, a dynamic, object-oriented programming language.

Kay once claimed he "could define the world's most powerful computer language in a single page of code."[9] At PARC, he was challenged to do so, and his past track record indicated that the results could be surprising: earlier, while working for ARPA and the University of Utah, he had developed the new programming approach known as dynamic object-oriented programming; served on a team that developed continuous tone 3D graphics; elaborated innovative computer design ideas such as those in the FLEX machine and another early concept for a desktop computer; and contributed to the design of the ARPAnet.

After joining PARC in 1972, he continued his work on Smalltalk and he also completed the design of a laptop computer that he called the *DynaBook*. He also participated in the development of Ethernet, laser printing, and client–server software architecture. He is well known for his contributions to object-oriented programming and user interface design.

Eventually, Kay left PARC to take influential positions at Atari, Apple, and Disney. He is now the head of Viewpoints Research Institute, a nonprofit organization established to advance ideas in education and personal computing.

In addition to Kay, Robert Taylor was successful in attracting other stars into his orbit. He persuaded Xerox to employ several of his colleagues from the University of Utah, and he also recruited scientists from Carnegie-Mellon and MIT. But Taylor most wanted to hire Butler Lampson.

BUTLER LAMPSON

Born in Washington DC in 1943 to Foreign Service parents, Lampson grew up in Turkey and Germany before returning to the United States. After completing high school, he went on to study physics and computer science, receiving his B.A. degree in physics from Harvard University in 1964, and his Ph.D. in electrical engineering and computer science from the University of California at Berkeley in 1967.

Lampson gradually shifted from his studies in physics to his focus on computer science. While at Harvard, he did some computer programming for faculty in the physics department. When he moved on to UC Berkeley, his original intent was to continue in his study of physics; however, he became sidetracked when he was introduced to the concept of time sharing.

According to Lampson, "When I went to Berkeley to continue studying physics, a very interesting computer research project was going on, but it was well concealed. I found out about it from a friend at a computer conference I attended in San Francisco. He asked me how this project was doing. When I said I'd never heard of it, he told me which unmarked door to go through to

[9]A. C. Kay, "The Early History of Smalltalk," in *History of Programming Languages-II*, ACM Press, New York, 1996.

find it."[10] That secret door led Lampson to Project Genie, the ARPA supported time sharing effort. Lampson joined the team and switched his major from physics to electrical engineering.

Having joined PARC in 1970 as a founding member in the Computer Science Laboratory (CSL), he soon made important and visionary contributions to the development of the Alto. He expressed his early vision of a personal computer in a 1972 memo entitled "Why Alto?"[11] He contributed to a broad range of projects related to computer architecture, networking, raster printers, and tablet computers.

By the early 1980s, Lampson left PARC and went to Digital Equipment Corporation (DEC). He subsequently moved to Microsoft Research, where he is a Microsoft Corporation Technical Fellow; he also is an adjunct professor of computer science and electrical engineering at MIT.

CHARLES (CHUCK) THACKER

Another brilliant scientist recruited by Robert Taylor was Chuck Thacker, who, like Lampson, was just turning 27 when he joined PARC. Thacker grew up in Los Angeles, the son of an electrical engineer. Immediately following high school, he was unsuccessful in his attempts at university studies at Cal Tech, and then UCLA. By 1963, at the age of 20, he found himself ready to start over again at the University of California at Berkeley. A year later he got married and settled into his work toward his B.S. degree in physics, which he received in 1967.

In 1968, Thacker became a junior engineer at the ARPA-sponsored Project Genie, intending to remain only a short time before returning to graduate school for more academic training in physics. Instead, it became a pathway to his future in computer science.

After joining PARC in 1970, he went on to become the project leader and chief designer of the Xerox Alto personal computer system. As chief designer, he took a hands-on approach writing portions of its microcode, constructing the first few Altos, and supervising the production of the rest. He also contributed to the invention of the Ethernet and several other innovations, such as the laser printer.

In planning the Alto project, Thacker, along with Lampson, realized that they would have to make a system that was both cheaper and better than minicomputers. Otherwise, replacing the more expensive multiuser systems based on time sharing with dedicated personal computers would not be a viable option.

[10]Biographic sketch and interview of Butler Lampson. http://research.microsoft.com/Lampson/37a-ProgAtWork/37a-ProgAtWork.htm.

[11]B. Lampson, XEROX Inter-Office Memorandum "Why Alto," December 19, 1972. http://www.digibarn.com/friends/butler-lampson/index.html.

To Thacker and his team, a personal computer would need to be as easy to operate as a typewriter. Only a state-of-the-art system could provide the desired functionality, but at the same time, only a stripped down system could be affordable. Thacker's team introduced an approach called *bitmapping*, which allowed a one-to-one correspondence between the bits in the computer's memory and the picture elements on the graphic display. This measure would prove to help in controlling the costs of the embryonic personal computer system.

Computers consist of four chief functional components: input, central processor, memory, and output. Data and instructions are transmitted from an input device such as a keyboard to memory. The central processor retrieves data and instructions from memory, and it executes the required steps. It then dispatches the results to an output device such as a screen or a printer.

This scheme has an added complexity. The central processor itself consists of two subunits. Its *arithmetic and logic unit* manipulates the data to produce computing results; its *control unit* keeps order throughout the system, much as an air traffic controller directs takeoffs and landings to prevent collisions.

Thacker realized that sharing of a processor's cycles between input and output could reduce the cost of an Alto. The dilemma was how to cut back on hardware without sacrificing features like the bitmap display that required access to powerful electronics—in other words, how to subtract circuitry while adding capability. Thacker later indicated that the idea had come to him suddenly; it was an instantaneous moment of inspiration.

His innovation, called *multitasking*, in reality turned one processor into many. He wired the control unit of the Alto's central processor to take its instructions from up to sixteen different sources, or *tasks*, instead of the usual one. Among these tasks were the bitmap display, the mouse, the disk drive, the communications subsystem, and the user's program. The tasks were assigned priorities: if two or more of them signaled a request, the one with the highest rank took possession of the processor. When the display was in control, it was the display's processor; when the disk drive had precedence, it was the disk drive's processor; when the mouse took charge, it was the mouse's processor, and so on.

Multitasking provided more functionality at a lower cost. Priority control of the Alto's powerful central processor meant the computer's input and output facilities could perform sophisticated feats without their own circuitry. Total system hardware requirements dropped by a factor of 10. The parts bill for an Alto ran just over $10,000, about 60% less than the cost for a minicomputer.

The multitasking did make the Alto slower, however. The bit-mapped display controlled the processor two-thirds of the time, leaving the rest of the system just one out of every three clock cycles to complete its work. Therefore, instructions and data took three times longer than normal to complete. The delays, however, were measured in microseconds. Furthermore, unlike time sharing, the speed of the Alto was thoroughly predictable.

After 13 years at PARC, Thacker left Xerox and went to work for DEC Systems Research Center. Subsequently he joined Microsoft in 1997, where he has continued to build his reputation as an innovator in personal computing.

PERSONAL COMPUTING

PARC was originally set up as a commercial research lab to support new product opportunities for its parent corporation, Xerox. From the beginning, however, PARC functioned more like an academic institute or an independent national laboratory for computer research than a corporate research and development center, thanks in large part to the autonomy and independence that Xerox allowed. As a result, great creativity was enabled, but at the same time, many of its innovations ended up in the public domain or in the hands of other corporations either through agreements or because of employees leaving to create their own start-up companies. Ultimately, the technical innovations and the people who invented them would begin numerous start-up firms and inspire the imitation of others, including Apple Computer Inc. and 3Com Corporation. However, primarily as a result of the decisions of Xerox management, the Alto itself failed to become a success in the commercial market.

Besides inventing particular pieces of hardware or software to support an overall vision, PARC was a place for thinkers who were able to expand the vision or head off in new directions if appropriate. Many computer companies are linked directly to that research center in California; being close to Stanford University, PARC attracted many talented researchers and served as a cornerstone of today's Silicon Valley.

Shortly after its formation, the PARC team, under Taylor's direction, decided to embark on a project to develop a display-based, interactive PC, destined to become the Alto. Lampson and Thacker had seen a useful interface in the research of Douglas Engelbart. They worked hard to incorporate the ideas of Engelbart for a graphical user interface, a mouse, a local disk drive, and a keyboard. The mouse was designed as an analog device consisting of a housing, steel wheels, and buttons. The motion of the wheels controlled the cursor, the highlighted marker on a graphical display screen. By moving the mouse over a flat surface, the curser on the display also moved to allow pointing and selection of graphic objects.

Many of the ideas incorporated into the Alto were not necessarily new, but nevertheless, much of the technology had advanced to the point that synthesizing the pieces into a new system had become viable. Furthermore, by the late 1960s, the falling prices of semiconductor memory and processor chips could be projected using Moore's Law, and the cost viability of the personal computer could be confidently predicted. Thus the stage was set for PARC to embark on its ambitious project to bring together all the technology pieces needed to pave the way for the personal computer.

In fact, during the decade long period of development of the Alto, PARC introduced a spectacular array of technology innovations that have had far-reaching impacts. These include Ethernet, Smalltalk, and the first page-description word processing language.

THE XEROX ALTO

PARC offered its scientists the freedom to investigate a broad range of creative projects. It had attracted the top computer science researchers in the country and the atmosphere was dynamic and exciting. PARC researchers believed they were inventing the future of computing.

One of their first inventions was the laser printer, which later generated revenues amounting to $1 billion per year for Xerox. But the printer required a more graphical approach for a computer to prepare documents. Since the computers of the time were not able to easily accommodate the laser printer, PARC was motivated to create new, more compatible computing technologies, and so the researchers began work on their own computer system, the Alto, in 1972, which they were able to demonstrate a year later.

The Alto was intended to become the first general purpose computer system designed for the individual user. The idea at PARC was to create a real-time, interactive machine. The first Alto was completed in April 1973 and this first-of-a-kind unit provided the ability to perform interactive computing. The graphical display approach taken for the system allowed a What You See Is What You Get (WYSIWYG) approach to computer output, a particularly important innovation considering the connection of this technology to the Xerox laser printing technology.

One of the Alto's most striking features was its display, which was the same size and orientation as a printed page. The system operated on the idea that each pixel (picture element) of the display could be independently activated. It had a keyboard and a mouse, which used the bitmap image concept in their operation.

Computer hardware without the software to run it would be like a car without gasoline. The concurrent development of software for the Alto was another priority at PARC. The first software for the Alto was file management software. Then, a graphics-oriented word processor, called Bravo, was developed by Butler Lampson and Charles Simonyi. Bravo was particularly innovative in that it demonstrated the WYSIWYG concept to its greatest extent at the time. Text was presented on the monitor, where it could be modified on-screen with such formatting characteristics as typeface, font size, bolding, underlining, and italics, while the page image on the screen would exactly match the printed product on the selected laser printer. Bravo was used throughout Xerox, shaking out the technology and causing the news of this important advance to become widely disseminated.

Interestingly, software developer Simonyi would subsequently join Microsoft and recreate his work by writing Word for DOS. The experience in software development at PARC brought about the realization that what was needed was a consistent user interface; to achieve this they chose Smalltalk, the software for the first modern GUI.

Taken together, GUI, Smalltalk, and the Alto hardware system were the ingredients for a modern personal computer. The Alto system had the capability of networking through its Ethernet capability and could be used to send email, another key innovation that turned out to be even more valuable than originally anticipated. The PARC research staff was enthusiastic about getting the Alto system to market, but they would have significant obstacles to overcome, primarily from Xerox management.

Following the completion of the first Alto, the Alto I, subsequent versions (i.e., the Alto II and Alto III) were introduced, still roughly three years prior to the release of the IBM PC. Nevertheless, the Alto III was a machine clearly superior to the IBM PC, which was packaged with a monochrome display and without a mouse. Although staff members at PARC urged the commercial release of the Alto, Xerox management chose instead to go forward with a much less forward-thinking nonprogrammable word processor known as the Xerox 850. Xerox would not attempt to compete in the PC market until much later when, in 1981, eight years after the invention of the Alto, it attempted to commercialize the computer known as the Xerox 8010 Star.

While research scientists at PARC pioneered GUI technology and developed the revolutionary Smalltalk system software that enabled Alto to work like an early version of Windows, PARC's visionary work would inspire others as well; for example, Steve Jobs, having toured PARC in 1979, was inspired to develop the Macintosh for Apple Computer Corporation, while Bill Gates, after examining a prototype Macintosh, was inspired to develop Windows.

On his visits to PARC, Jobs was so overwhelmed by the graphical user interface that he changed the design of the Apple Lisa system, already under way, so that it would be GUI based. In his enthusiasm over this particular area of innovation, Jobs apparently failed to recognize the future of the Smalltalk object-oriented programming and Ethernet technologies.

APPLE COMPUTER

On April 1, 1976, two college dropouts, Steve Jobs and Stephen Wozniak, founded the Apple Computer Company. They began operating out of a garage, building the Apple I, which some claim to be the first personal computer to be sold as a fully assembled package. Fearing disaster, the third cofounder—unfortunately—sold his 10% stake in the partnership for $800 less than two weeks later.

In the early 1970s, before the introduction of the Apple I, the personal computing products available on the market had limited appeal. They were generally sold by small electronics firms and individual hobbyists through clubs. In many ways, Wozniak's Apple I still typified the early merchandise. It consisted of an unpackaged circuit board wired by Wozniak so that a purchaser could hook it up to a power supply. Within a few years, however, astonishing advances in integrated circuitry provided the critical raw materials needed. And programmers began writing software to make the machines appealing to people.

In 1977, Wozniak and Jobs introduced the Apple II. In stark contrast to the Apple I, fundamentally a kit computer with limited appeal though creatively priced at $666, the $1298 Apple II is considered by many to be the first personal computer designed for the mass market. Market appeal came from its attractive physical design, and the fact that it came fully assembled with a standard keyboard, integrated power supply, and color graphics capability.

From its start, Apple Computer had the necessary magic, in terms of both faith and hustle, to become a successful venture. It was a classic American business story featuring two high school graduates with little money, no training in business or economics, and big dreams. Wozniak designed and built the machines; Jobs provided the faith and the hustle.

It is interesting to note that in 1979 Jobs visited Xerox PARC and had a chance to see the future of computing. PARC and Apple set up an agreement in which Apple provided Xerox 100,000 private shares of stock at the price of $10.50 per share, in exchange for the alliance between the two organizations that allowed Apple engineers in on some of the PARC secrets related to GUI technology. After the Apple engineers played with Xerox's GUI for a day, they went back to Apple and started to work on implementing the ideas they saw.

Software is always an issue for new computing systems. At first, many Apple programmers developed software primarily for games. But by 1979, the technology began to mature, and applications such as database management, word processing, and spreadsheets all had been introduced. With the emergence of these applications, large numbers of people realized that the small computers could help them manage information more productively. As a result, the personal computing market measured revenues in the billions of dollars by 1981.

Apple has always prided itself on its work atmosphere as one that provides a creative, freewheeling environment to promote fresh ideas and innovation. The contrast between this image of a corporation and that of Apple's arch competitor IBM was addressed in the previously mentioned Apple advertisement in the year 1984. In George Orwell's novel *1984*,[12] he suggested that the computer would allow power-hungry people to rule the world. Apple held out the opposite promise. Apple promoted its alternative image in its commercial

[12]G. Orwell, *1984*, 1st World Library, July 2005.

shown on the 1984 Super Bowl broadcast. The ad begins with gray-clad ideological slaves marching in lockstep toward a great hall, where they take instruction from a larger-than-life image projected on a screen. In the midst of this lifeless, impersonal scene, an athletic woman dressed in bright colors wields a sledgehammer, smashing the image. In this landmark of advertising history, Apple not only expressed its alternative vision but also drew the battle line clearly between itself and IBM.

In 1985, President Ronald Reagan awarded both Wozniak and Jobs the National Medal of Technology, the highest honor bestowed on America's leading innovators, for their achievements at Apple Computer and their contributions in bringing the power of personal computing to the general public. In speaking of his experience, Jobs said: "If we could transform high-cost, complicated computer technology into a product that was very low-cost, easy to use and highly reliable, then it had the potential to change people's lives and even change the world."[13]

The success of the Macintosh put Apple Computer on the map. It also resulted in Microsoft recognizing the importance of GUI to future sales. Eventually, the personal relationship between Jobs and Bill Gates led to a period of cooperation, where Microsoft learned the basics of GUI technology, allowing Microsoft to begin its own project: Windows.

The reincarnation of the Xerox Alto's GUI first in Macintosh and then in Windows is a testament to its fundamental value.

IBM PC

IBM dominated the large-scale computer industry in 1970, holding more than 70% of the market. IBM had several advantages. First, it had a nationwide sales force already serving most businesses looking to buy computers. Second, it already understood how to automate recordkeeping. Third, IBM leased instead of sold its equipment. Long after the machinery had paid for itself, lease checks kept rolling in. Things were about to change, however, in 1972, with Intel Corporation's introduction of the 8008, the first 8-bit microprocessor. But, as always, software had to catch up with the new hardware before major advances could be achieved. Bill Gates, Paul Allen, Gary Kildall, and others were players in the software development events leading to the introduction and remarkable success of the IBM PC.

With the popularization of personal computing in 1975, Paul Allen and Bill Gates seized the opportunity on the software front and developed BASIC for the Altair 8800. This quiet event resulted in the birth of Microsoft, an event that would have major and long-lasting implications for the personal computer industry. Microsoft was created when Gates and Allen, childhood friends,

[13]The Spirit of American Innovation, "Stephen P. Jobs and Stephen Wozniak: The Personal Computer Is Born." `http://www.thetech.org/nmot/detail.cfm?ID=17&STORY=3&`.

formed a corporation to write software for personal computers, and they achieved their first success with the Altair 8800. The two were acting on a vision that they had of a desktop computer in every office and home, a vision that few others were able to see at the time, but one that has been more than fulfilled by the present time. Gates and Allen went on to write versions of BASIC for other microcomputers as they came to market.

Meanwhile, Gary Kildall invented the first microcomputer operating system. In the early 1970s, Kildall taught computer science at the Naval Postgraduate School in Monterey, California. He wrote compilers, software tools that take entire programs written in a high-level language like FORTRAN or Pascal and translate them into object code, coding that can readily be converted into forms that can be executed directly by computers.

By 1974, Intel had introduced the 8080 line of microprocessors, and they hired Gary Kildall to write software to emulate the 8080 on a DEC time-sharing minicomputer system for software development purposes.

Kildall was bothered by the prospect of driving the 80 miles from his home in Pacific Grove to use the Intel minicomputer in Silicon Valley. While he could have used a remote teletype terminal at home, that approach would have been very slow. So he decided to develop software directly on the 8080 processor, bypassing the time-sharing system. To do this, he developed the operating system called CP/M (Control Program/Monitor).The operating system controls the flow of data between a computer and its long-term storage system. It also controls access to system memory and keeps those bits of data that are thrashing around the microprocessor from colliding into each other.

By the time he'd finished writing it, Intel was no longer interested in CP/M. Nevertheless, Kildall, along with his colleague John Torode, adapted CP/M to other systems. And they modified CP/M by identifying the parts that interfaced with each new hardware controller and combining them into a separate software module, called the Basic Input/Output System, or BIOS. With all the hardware-dependent parts of CP/M concentrated in the BIOS, it became a relatively easy job to adapt the operating system to many different Intel-based microcomputers by modifying just the BIOS.

With CP/M and the invention of the BIOS, Kildall had effectively defined the software capabilities of the microcomputer. He started a company called Digital Research. Digital Research was slow in adding software language development to its operating system business. It was also slow in updating its core operating system and extending it into the new world of 16-bit microprocessors that came along after 1980.

In August 1981, IBM introduced its new revolution in a box, the "personal computer," which was packaged as a complete system including the brand new operating system from Microsoft and a 16-bit computer operating system called MS-DOS 1.0. The cost of the entire package was $1565, and within two years IBM's market share eclipsed that of Apple.

In 1980, Bill Gates's Microsoft and IBM met to discuss the technology of personal computers and the role of Microsoft's software products. As a result of the meeting, IBM adopted the idea of including BASIC in the computer's firmware, and Microsoft, having already produced several versions of BASIC for other computers, agreed to produce a version for IBM.

In terms of an operating system, Microsoft had suggested the use of CP/M. Kildall's successful venture had resulted in the sales of over 600,000 copies of this operating system to date, and it was considered to be the operating system standard of the day.

When IBM found itself unable to contact Kildall directly, they came back to Microsoft and entered into a new contract for operating system software. The result was the creation of the Microsoft Disk Operating System (MS-DOS), written to be similar to CP/M, but different enough to be considered legal. As part of the deal, Microsoft convinced IBM to allow it to retain the rights to market MS-DOS separately from the IBM PC project. In the end, Microsoft was to enjoy huge profits from the licensing of MS-DOS.

By this time, IBM's name had become synonymous with the PC, and it was so closely identified with personal computing that the avalanche of PC knock-offs became commonly known as IBM PC clones.

The first Microsoft GUI was called Windows. It wasn't very powerful and, as a result, was not particularly well accepted. Follow-on versions of Windows were written, but the first really popular one was Version 3.0, released in 1990. It capitalized on the improved graphics available on PCs by this time, and on the increased capabilities of the new Intel 80386 microprocessor, which would enable true multitasking within Windows applications. This made it much more efficient and much more reliable when running more than one software application at a time. It would even allow older MS-DOS-based software to run, providing a level of upward compatibility of software. It was Windows 3 that moved the IBM PC into serious competition with the capabilities of the Apple Macintosh.

Windows 95 was released in 1995 for use on the new IBM PCs that included a mouse, Ethernet networking, object-oriented programming, a laser printer, and a WYSIWYG word processor.

Does that sound familiar? It should! It's very similar to the Xerox Alto with its complete set of inventions from Xerox PARC.

PATTERNS OF DISCOVERY

The problem that the PARC, Apple, and IBM teams were trying to solve was to provide easy interactive computing for individuals. It required innovations including CPU chips and memory power, operating system, mouse, network, printer, and user interface.

PARC developed a research laboratory and expert team to address these issues as an integrated problem. The result was the world's first personal computer and networking system.

The Xerox PARC Alto may have been the first and best proof of principle invention in this area, but it was a commercial failure that left the door open for innovators such as Apple and IBM. Both Apple II and the IBM PC overshadowed the Alto initially, but the IBM PC eventually won out over the Apple through its open architecture.

While the PARC team used the Proof of Principle Pattern to accomplish its invention of GUI, Ethernet, laser printing, and the mouse, the pattern of success of both Apple II and the IBM PC was the 1% Inspiration and 99% Perspiration Pattern. Clearly, achieving commercial success was different from scientific success.

FORECAST FOR CONNECTING CHIPS

In 1975, fewer than 50,000 personal computers were sold. They had a value of about $60 million. From this limited start, the PC industry grew to sales exceeding 128 million units per year and revenues surpassing $225 billion by 2002. While the PC industry will continue to grow, we can expect that it will be at slower rates.

The Asia Pacific region is becoming the largest area for PC use, while in the United States, the growth rate of PC usage is likely to decline. Europe also has large PC usage levels, while the rest of the world is further behind in per capita PC sales and use, and therefore has greater room for continued growth.

Desktop PCs are the predominant form of computer devices in current use, but PC servers and mobile PCs, including laptops, tablet PCs, and wearable PCs, represent a growing market. In addition, personal digital assistants (PDAs), including crossover devices such as cell phones, are rapidly growing in use.

We can expect the PC industry to continue to connect chips and exploit innovation and development.

Discoveries Requiring Inspiration and Perspiration

It is likely that, while the PC industry will continue to grow, this growth will be at slower rates, simply because the sheer size of the industry limits its growth rate. Nevertheless, some estimates indicate that annual worldwide sales of PCs may grow by over 80% in the next few years and could surpass 300 million units by 2009.

The Internet is a key factor in the growth and development of the PC and computer industry. Over the next decade, however, one can expect that new technology developments will be pursued in the areas of consumer electronics with microprocessor capabilities (i.e., information appliances) and mobile computing devices (i.e., PDAs, Smartphones, ultraportable computers, and the like).

Information appliances have been identified as part of the emergence of the post-PC era. Part of this stems from the fact that microprocessors are finding themselves in more and more consumer products, from automobiles to greeting cards. With wireless networking technology expanding its reach, a trend toward ubiquitous computing appears to be emerging. Information appliances represent a growth area and an area ripe for innovation. As part of the growth in the use of information appliances, especially those that are connected via the World Wide Web, there will be a corresponding need for server hardware and software to support the resulting needs for infrastructure enhancements.

Anticipated new developments in the computer market include multi-computer home networking, multimedia PCs, portable systems including the tablet PC, and smaller systems with enhanced capabilities such as voice recognition.

By 2015, merely allowing for chips already in the Moore's Law pipeline and following the 1% Inspiration and 99% Perspiration Pattern, we could expect new supercomputer capabilities to continue to grow rapidly. An early indication of this can be seen in IBM's announcement in 2005 that it had doubled the performance of the world's fastest computer, named Blue Gene/L, from 136.8 trillion calculations per second (teraflops) to 280.6 teraflops.[14]

It is the quest for ubiquitous computing that is one of the most potent drivers of computer-related innovation and development. Although many related technology advances are currently under way (e.g., wireless networking, mobile hot spots, advanced speech recognition and language translation, smart cell phone technology), the integration of these advances into a capability to provide seamless mobile access to computer and communications resources any time and anywhere is yet to be achieved. All the component technologies appear to exist, and the inspiration/perspiration pattern can be expected to result in successful convergence.

Discoveries Requiring New Proof of Principle

If we allow for discoveries based on new Proof of Principle Patterns, then we could expect the development of medical applications that might include special purpose microfluidic chips with capabilities such as blood analysis or gene sequencing; the capability for electronic medical diagnostics allowing self-certification for prescriptions; and remote telemedical consultations or the electronic office call. There could also be applications of advanced computational capabilities to determine the genetic links to human diseases.

In addition, there may be new developments based on new structures and materials for electronic devices along with mechanical intelligence using micro- (MEMS) and nano- (NEMS) electromechanical systems.

[14] "Imagining the Internet: A History and Forecast," Elon University/Pew Internet Project. `http://www.elon.edu/predictions/`.

Discoveries Based on New Serendipity Pattern That Could Be Developed

Exotic new computing technologies require serendipitous discoveries to make them successful. These might include quantum and DNA computing that uses DNA and molecular biology instead of the silicon-based computer technologies.

One of the businesses that could result from DNA computing would be a genetic analysis service. Researchers from Columbia University Medical Center in New York, the University of New Mexico at Albuquerque, and the National Science Foundation say a DNA-based computer could lead to faster, more accurate tests for diagnosing West Nile virus and bird flu.[15]

[15]"DNA Computing Targets West Nile Virus, Other Deadly Diseases," *Physorg.com*, October 16, 2006. http://www.physorg.com/news80230272.html.

4

Connecting Processes

There are two ways of constructing a software design; one way is to make it so simple that there are obviously no deficiencies, and the other way is to make it so complicated that there are no obvious deficiencies. The first method is far more difficult.
—C. A. R. Hoare[1]

Art is something that visually stimulates our thoughts and emotions. As a result, we prize and admire it. We call the creative individual who portrays our world in this type of imagery an artist. If the depiction is successful, the artist is esteemed and his/her work is shared and valued.

But art can take many forms. It can even take the form of symbols, such as in a computer program. And this creative individual—a programming artist—portrays the world through logic symbols: his/her work, if successful, is also respected, shared, and valued.

Consider the different schools of art. Realism, Impressionism, Abstract Art, and Modern Art were styles developed during succeeding periods that reflect changing social ideals and values. Within each period, the artist represented the world through different types of imagery. These successive periods and their corresponding styles could be considered to represent successive generations in the evolution of artistic expression.

Similarly, computer programming has passed through successive generations in its evolution. In the first generation of software development, computer instructions were written in what is called *first generation languages* (1GLs); these were primitive machine languages in which the computer

[1]C.A.R. Hoare, "The Emperor's Old Clothes," The 1980 ACM Turing Award Lecture, *Communications of the ACM* 24 (2): February 1981.

Connections: Patterns of Discovery By H. Peter Alesso and Craig F. Smith
Copyright © 2008 John Wiley & Sons, Inc.

instructions were written explicitly in 1's and 0's. In succeeding generations, higher order languages replaced the use of these binary numbers with higher level symbols so that more complex instructions could be represented in ways that were much easier for humans to comprehend. Approaches using structured and object-oriented programming were reflected in follow-on generations and represented changes in technology and performance. Within each generation, the programmer represented directions to the machine through different types of symbols and concepts as a means of *connecting processes*. Each generation attempted to provide the maximum in expressive power while remaining as simple as possible.

Over the years, many programmers have applied the most current methods to create computer instructions in routine and prescriptive ways, but only a select group of programmers can be considered to be artists—programmers who have developed new software concepts, approaches, capabilities, and standards. A few of the outstanding programming artists from different generations include John von Neumann—who developed early ideas for software architecture; Claude Shannon—who established the basic concepts of programming; C. A. R. Hoare—who established fundamentals of programming logic; Charles Simonyi—who developed novel approaches to graphics-based software; Bill Gates—who found proprietary dominance in software; and Linus Torvalds—who established open operating standards.

Perplexing questions that emerge from considering the evolution of software from its succeeding generations are: Why hasn't software productivity mirrored the billion-fold improvement in computer hardware observed in Moore's Law? Should future software be designed and developed from the top–down (through an authoritative structured organization) or from the bottom–up (through self-organizing software and the use of open source components)?

In this chapter, we tell the story of software, its evolution through successive generations, its programming artists, and the future directions it may take in addressing these perplexing questions.

THE SOFTWARE STORY

The modern computer serves a myriad of different purposes, from word processing and advanced scientific calculations to videogames and graphics. Since the computer is basically a simple device whose principles of operation (input, output, memory, and processing) haven't really changed over the years, how can it behave in so many different and complex ways? The key is in the software.

Computers perform complex tasks by following simple instructions. The instructions can be grouped into combinations that we can call processes. These processes are executed by the computer's processing units, called *processors* or *microprocessors*. The wonderful and surprisingly complex behavior of computers comes from the connection of these processes in new and different ways. The end result is what we call a program, or collectively, software, so called because it is fundamentally information and lacks the physical characteristics of hardware.

Computer programs are designed to process input data and produce useful results. It is said that the first program was produced by Ada Lovelace, the daughter of English poet Lord Byron, in 1843. In notes to her written summary of the general purpose mechanical computer designed by philosopher and engineer Charles Babbage, called the *Analytical Engine*, she described the steps for calculating a number sequence known as the *Bernoulli numbers* using the Engine. Many credit her with creating the world's first computer program and, although Babbage's Engine never became operational, software had its first artist.

From this beginning, it was clear that computing machines, such as the Analytical Engine, would be simple devices capable of complex behavior as determined by the machine's program, or set of instructions. Once computers were introduced into widespread use, the methods and techniques of programming or software development began a rapid evolution that paralleled the evolution of computer hardware.

Early computer systems were programmed by very tedious methods of providing the needed input. Prior to the 1950s, the first computers were programmed by physically changing wires, dials, and switches. As a result, the programming of the first computers was not an easy task. It was prone to mistakes. Eventually, information carriers such as punched cards, first introduced in 1890 to assist in the completion of the U.S. census, were used to tell machines what, how, and when to do something. This greatly reduced the chance of errors and allowed the development of progressively more complex programs.

The EDVAC and the UNIVAC-I were two of the first generation of computers, computers based on wired circuits, vacuum tubes, and punched cards. They took advantage of the idea of random access memory (RAM), which allowed a program to be stored in the computer's internal memory rather than being executed step-by-step from externally applied commands. These early computers, operating in the early 1950s, had an internal RAM capacity of 1000 words, which, though miniscule even in comparison to today's simplest personal computers, represented an important leap forward at the time. In terms of size, these computers were much smaller than ENIAC, the first large-scale digital computer capable of general purpose programming. They needed frequent maintenance and reached only 80% reliability. These first generation computers were programmed directly in machine language, instructions encoded in numeric code that can be directly interpreted and acted upon by the computer hardware.

Ultimately, computing machines store both data and program instructions in strings of binary digits (bits), coded as 0's and 1's. In machine language programming, the programmer had to encode instructions in the same numeric form. An error in a single zero or one could mean disaster, and with machine language programs, all coding instructions had to be absolutely in the right place and in the right order in memory.

The outstanding mathematician and computer science pioneer John von Neumann, the next of our programming artists, took one of the first critical steps toward formalizing the art of programming when he developed two key programming concepts: the *shared-program technique* and the concept of *conditional control transfer.*

JOHN VON NEUMANN

John von Neumann was born December 28, 1903 in Budapest in what was then Austria-Hungary. Von Neumann was a child prodigy who demonstrated advanced skills in mathematics and language even at a very early age. It is said that he could speak fluent ancient Greek and complete long division in his head by the age of six. He mastered calculus while in primary school, and before he was a teenager, he was familiar with complex mathematical theories normally addressed in university work at the graduate level. He was not narrowly focused in his interests, however. He was deeply interested and well informed in history, having read the complete 44-volume *Universal History* at the age of eight.

One of the most brilliant scientists of the 20th century, he would become well known as a leader in several different fields, including mathematics, nuclear and quantum physics, economics, neurology, and computer science. Among his great accomplishments was his role in conceptualizing and defining the operating processes of the modern digital computer, thereby creating the foundation for computer programming and software development.

In fact, it is now common to refer to the *von Neumann architecture* as the model of the stored-program computer, a construct that is comprised of a general purpose computing system (i.e., a universal Turing machine) that uses internal memory to hold both the program and the data that is to be processed. And the term *von Neumann programming language* refers to any high level language that can be used to encode computer instructions consistent with this concept of computing architecture.

Upon completion of high school in Budapest, von Neumann went to the University of Berlin and then the prestigious Eidgennossische Technische Hochschule (ETH) in Zurich to study chemical engineering. While studying chemical engineering abroad, he concurrently enrolled in the Eötvös Loránd University (ELTE), which is often referred to as the University of Budapest, to pursue his Ph.D. in advanced mathematics. By 1926, he was awarded his undergraduate degree from ETH and his Ph.D. with highest honors in mathematics from the University of Budapest. Following his university studies, he took a position at the University of Göttingen in Germany, where he worked on the mathematical theory of quantum mechanics.

By the end of the decade of the 1920s, von Newmann was a rising star in the scientific community and was offered a lectureship at Princeton University. At about this time, he married his childhood sweetheart, and his honeymoon was the sea cruise that transported the two of them from Europe to New York. In 1933, he was invited to become an inaugural member of the Institute for Advanced Study at Princeton, an opportunity that allowed him to work with other luminaries of physics and mathematics, such as Albert Einstein and Kurt Gödel.

Between 1936 and 1938, von Neumann interacted closely with the enigmatic Alan Turing, sometimes called the father of modern computer science, who came to Princeton University to complete his Ph.D. in mathematics. This period of interaction began in 1936 just after Turing had published a landmark paper describing the fundamentals of logical design for a "universal computing machine." Von Neumann was strongly influenced by his interaction with Turing and tried to get him to stay on at the Institute for Advanced Study; Turing decided instead to return to Cambridge, and a short time later joined the British war effort at their code-breaking laboratory at Bletchley Park. Stimulated by the interaction, however, von Neumann continued to develop his ideas on logic and computing machines until 1945, when he introduced two important concepts that would set the foundation for the field of computer programming.[2]

The first of these concepts is known as the *shared-program technique*. This idea proposes that computer hardware should be simple, flexible, and programmable; that is, computers should not be designed to need to be hardwired or set up differently for each problem, but rather they should be capable of acting on a series of instructions that in turn would reprogram the hardware. This concept ensured that computer hardware design would be simple, but software languages would need to have the full range of capabilities to take advantage of computers designed as universal computing machines.

The second concept was called *conditional control transfer*. It presented the idea of subroutines, or small blocks of code that could define reusable sequences of instructions to perform processes that could be combined in any order. It also identified the necessity for computer programs to allow branching based on logical statements. Thus a program could proceed down one path if a logic expression was determined to be in one state (e.g., true), and proceed down a different path if the logic expression produced the opposite response (i.e., false). This led to the idea of conditional branching statements known as IF-THEN statements, and looping commands known as FOR statements. Conditional control transfer gave rise to the idea of reusable blocks of code that could be kept in libraries and retrieved whenever needed.

As a result of these two concepts, the fundamental ideas of programming were established. Software languages would allow sequencing, branching, and

[2]A. Fergusen, "The History of Computer Programming Languages." `http://www.princeton.edu/~ferguson/adw/programming_languages.shtml`.

looping in a hierarchical organization. These were the basic building blocks of all future programming languages.

When John von Neumann died in February 1957, he left behind a partially complete manuscript of a paper entitled *The Computer and the Brain*.[3] This was his last major creative project; it was a suitable legacy for this intellectual giant who was one of the most important pioneers of the era of the modern digital computer.

The stage having been set by the formulation of the fundamentals of software programming, much still remained to be done. In particular, the realization began to set in that computers could be used broadly in society, not just for number crunching but for much more complex tasks, including such things as logical reasoning and communications. Once again, the key to enabling these more complex functions lies not in the hardware design but in the software. Thus computing was ready to take another step into the new era when the American electrical engineer and mathematician Claude Shannon wrote his seminal paper "A Mathematical Theory of Communication." [4]

CLAUDE SHANNON

Claude Shannon, who has been referred to as the father of information theory, and who was a well-known pioneer in digital circuit design theory, prepared this paper in 1948 to lay out the theory of how binary logic could be used in computing and communications. His paper also introduced the concepts of information entropy and the binary digit as the fundamental element of communication. His work was instrumental in establishing the basic software concepts necessary for modern computing. It represented a big step forward in thinking.

Claude Shannon was born in Petoskey, Michigan in 1916, the son of a judge and a high school principal. He showed capability in mathematics from an early age and also showed promise as an engineer, building model airplanes, radio-controlled boats, and a telegraph line as a boy. He was encouraged in his mechanical interests by his grandfather, who was an inventor of mechanical devices and a farmer.

He attended public school in Gaylord, Michigan, graduating from high school in 1932. Then he enrolled at the University of Michigan, where he studied both mathematics and electrical engineering. Four years later, he was awarded bachelor degrees in both majors.

Immediately following his undergraduate studies, Shannon became a graduate student and research assistant at the Massachusetts Institute of Technology (MIT). As a graduate student at MIT, he had the opportunity to work with

[3]J. von Neumann, *The Computer and the Brain*, Yale University Press, New Haven, 1948.
[4]C. E. Shannon, "A Mathematical Theory of Communication," *The Bell System Technical Journal* 27: 379–423, 623–656, July and October 1948.

Vannevar Bush, one of the great science and technology leaders of the 20th century, and he spent considerable time and effort on Bush's Differential Analyzer, a 1930s era analog computer that used mechanical wheel-and-disk mechanisms to solve differential equations. His hands-on mechanical abilities served him well with his work on the Differential Analyzer, and he gained a reputation as a tinkerer. In his paper completed in 1938, "A Symbolic Analysis of Relay and Switching Circuits," which also formed the basis for his 1940 master's thesis of the same title,[5] he laid out his landmark theories on the relationship between symbolic logic and the operation of relay circuits, a topic that was to have great significance with the growing importance of the digital computer. By this time, Shannon was well on his way in his quest to quantify information in terms of physical properties. He modestly expressed this important endeavor in saying, "I just wondered how things were put together." [6]

In 1940, Shannon was awarded his doctorate degree in mathematics from MIT and he immediately went to the Institute for Advanced Study at Princeton University, having been selected to receive a prestigious postdoctoral National Research Fellowship. One year later, Shannon completed his fellowship and took a position at the Bell Labs, where he worked to achieve more efficient information transmitting methods and improved reliability for telephone communications.

In his work at Bell Labs, he introduced new ideas on the nature of information transfer, including the fundamental concept of information entropy. This theory was among the most important ideas presented in his previously mentioned 1948 paper, "A Mathematical Theory of Communication." Taking an analogy from the second law of thermodynamics, entropy can be considered to be the degree of randomness in any system, a characteristic that always increases in a system isolated from outside influence. Shannon used the analogy to demonstrate an equivalent principle that could apply to the information content in a message. He proved that in a communication that would normally be subject to noise, a signal could always be sent without error or distortion if it could be encoded in a way that permitted self-checking. In this case, signals could be received without error just as if there were no interference.

Shannon's work set the direction for the future of information processing. Crucial to Shannon's work was the realization that, just as the relay switches on the Differential Equalizer were always either open or closed—on or off—the state of such a machine could be represented by the state of its binary switches. Similarly, information in general could be represented by such binary encoding. And so he began to consider mathematical approaches to describe such binary states. He used the theory of the British mathematician and logician George Boole known as the *Boolean algebra of logic*, to establish a

[5]C. E. Shannon, *A Symbolic Analysis of Relay and Switching Circuits*, Master's thesis, Massachusetts Institute of Technology, 1940.

[6]"Claude Shannon, Father of Information Theory, Dies at 84," Bell Labs article, February 26, 2001. www.bell-labs.com/news/2001/february/26/1.html.

mathematical approach in which all equations were reduced to a binary system of zeros and ones.

In Boolean algebra, a statement of logic is assigned a value of one if it is true and a value of zero if it is false. Shannon visualized the on–off states of relay switches in terms of Boolean ones and zeros. He went on to realize that, by reducing information to this binary code of ones and zeros, information could be represented directly by on–off switches. Furthermore, by connecting such switches, analysis of mathematical equations could be carried out.

Having introduced the term *bit*, an abbreviation of the phrase *binary digit*, Shannon went on to consider the role of digital machines for computing, communications, and ultimately for the implementation of intelligent software. By combining mathematical theories with engineering principles, he was able to advance the state of technology and help set the stage for the impending revolution in digital computer and digital communication technologies. As a result, he became commonly known as the father of information theory.

Shannon became the victim of Alzheimer disease, which tragically ended his life in 2001. Thus in his later years, he was not aware of the profound changes that were taking place in society as the digital revolution took hold. His obituary[7] quoted his wife in saying that "he would have been bemused" by it all.

THE EVOLUTION OF PROGRAMMING LANGUAGES

Since the introduction of the digital computer, computer programming has evolved with the development of a series of progressively more complex and sophisticated approaches to programming. These approaches, known as the generations of programming languages, have been introduced in parallel with developments in the computer hardware they support. Generally, there are five generations that are recognized.

First generation languages (1GLs) are considered to be low level languages that consist of coding in machine language, the binary format that can be used directly by the computer processor. Second generation languages (2GLs) are also considered to be low level languages; however, they include the ability to replace strings of machine language bits by symbols that can then be processed by an assembler, a program designed to convert the symbols into machine language. Third generation languages (3GLs) are high level languages that allow a higher level of symbolic representation of actions desired by the programmer in forms that are easier for a human to understand than the simple mnemonics of the 2GLs. The fourth generation languages (4GLs) are languages that consist of higher level symbols, but tailored to the specific area of application such as business or scientific computing. Fifth generation lan-

[7]K. Coughlin, Claude Shannon obituary entitled "Bell Labs Digital Guru Dead at 84—Pioneer Scientist Led High-Tech Revolution," *The Star-Ledger*, February 27, 2001.

guages (5GLs) can contain visual tools to help develop a program. 5GLs can also be described as 4GLs with a knowledge base built in.

When computers first appeared, the efforts to operate them were slow and labor intensive. The hardware was directly rewired at each step in the process in order to perform a calculation. The introduction of the idea of stored-program computers in the early 1950s was a great step forward, but the programming for these first generation computers was primitive and tedious. Instructions for the computer, although they could be stored internally for automatic execution, nevertheless were written in machine language, long sequences of ones and zeros. The process was time consuming and error prone. Nevertheless, 1GL, the first generation of computer languages had begun.

It wasn't long before the idea of using mnemonic symbols to represent computer instructions was introduced, and 2GL, the second generation, was born in the mid-1950s. This approach allowed errors to be reduced and productivity to be increased as programmers could now use symbols that were much closer to the human thought process. However, it required that another step be added to "assemble" the 2GL code into the machine language that a computer could act on. This was done using an *assembler*, a software package that completed this direct translation of code on an instruction-by-instruction basis.

Toward the end of the 1950s, the third generation of programming languages (3GLs) began to be introduced. 3GL refers to the introduction of "natural language" interpreters and compilers. Both of these methods were similar to assemblers in many ways, principally because they serve the same function of converting symbolic code into machine language, but they also differed in some important aspects. The interpreter takes one line of high level code at a time, translates it into machine language, then executes that command without yet considering the next line of code. It consists of sequential translation and execution of high-level computer code, and because of this, its operation tends to be relatively slow.

In contrast, a compiler translates the entire code package into an intermediate form, called the object code. For execution, the object code modules, augmented by libraries of other software that can be invoked by reference, are linked together to provide a machine-executable package.

With a 3GL, there was no longer a need to work in abstract symbols. Instead, a programmer could work in a programming language that resembled natural human language. During the 1970s, a large number of 3GL "high level" languages were introduced with names such as Fortran, Pascal, Algol, PL/I, Basic, and C.

With the first three generations of programming languages, the emphasis was on the use of structured programming methods, where problems were viewed as being algorithmic in nature, subject to solution by sequential analytical steps. As we moved into the 1980s, there was a shift toward object-oriented programming and new ideas such as hyperlinking. Object-oriented programming languages incorporate the idea of data acting as a trigger for procedures.

Note that such ideas are not necessarily new; in the early 1970s, the language Smalltalk not only used this idea but extended it further, by providing for objects that encapsulate both local procedures and data, and exchanged messages among themselves, each message initiating appropriate behavior in the recipient object.

It was natural that a new generation of programming languages, 4GLs, would emerge to provide a higher level of abstraction in the programming tools and a simultaneous focusing of the applicability of the software to particular uses, while offering platform independence where possible. The objective of 4GLs is to offer software that will allow nonexpert users to solve their own problems while offering software that is more limited to the specific application; this is in contrast to 3GL software that tends to be more general purpose.

Finally, with the introduction of 5GLs, software began to include its own knowledge base. 5GLs represent a more complete development environment with the objective that the code be automatically generated. This way, code can be updated by modifying the 5GL statements, rather than by rewriting the entire code.

While it's interesting to review the evolution of software programming approaches, it is important to recognize that software has evolved not in a vacuum but right along with the computer hardware. In fact, in the early days, large computer vendors gave away the system's software just as system and application software are now included in a bundle with personal computers. As a result, in the early days of centralized computing, large organizations did their own programming and, although there were a few companies formed to provide custom software, by and large the idea of packaged software had not yet come about. It would take the advent of the personal computer to change that paradigm.

In addition, one of the most important developments for programming was the acceleration of the introduction of new computer systems. Once new computers began coming out almost every year or two, existing computers and their associated software became obsolete. Much of the work of programmers in the 1960s and 1970s was the continual rewriting of their programs to run on these new machines.

Even in the arena of large mainframe computers, however, some standardization began to emerge starting in the 1960s. 3GL compilers improved the platform independence of programs written in languages such as Pascal or Fortran, although by no means were they universally transferable. As the programming field stabilized, however, software became recognized as an increasingly valuable corporate asset. This stability led to the emergence of computer science as an academic discipline at many universities in the late 1960s, but software engineering was yet to be adopted as a recognized field of study.

When the number of computers began to rapidly expand in the 1960s, the demand for software grew accordingly. This finally provided opportunities for new companies to develop software. By 1965, at least 45 major software companies were operating in the United States, and some were quite large in size with more than a hundred programmers and annual revenues as much as $100 million.

Program application software at this time was being written primarily by single outstanding programmers who were assisted by several assistants who could modify, port, and convert the applications into other forms as needed. Other programmers would then imitate and copy useful features, but the era of the software artist was by now in full swing.

With the advent of the personal computer, things were about to accelerate. The 3GL Basic (Beginners All-purpose Symbolic Instruction Code) language was among the first programming languages adopted for general use of the nonspecialist. The popularity of this language was so widespread as personal computers came into common use that, until the end of the 1990s, almost no machine was sold without some version of a Basic interpreter. And the era of packaged software was also about to explode onto the scene.

As an interesting parallel development, the field of artificial intelligence (AI) came into existence in the 1950s. In 1958, the List Processing (or LISP) language was developed as part of the AI research at MIT. This language took a unique and forward-thinking approach by introducing the idea that the data used in this language is the list, comprised of a sequence of items enclosed by parentheses. Thus LISP programs are composed of sets of lists, and such programs have the unique ability of self-modification. This is a very important concept for the future development of software.

Other interesting AI languages include Prolog (Programming Logic), Smalltalk, Algol, and, to a lesser extent, Simula. But because the new concepts of AI software could not find direct implementation in commercial applications (with the significant exception of Smalltalk), their impacts on the evolution of programming have been limited to date.

In 1960, the British computer scientist Sir C. A. R. Hoare led the design of the first commercial compiler for the Algol60 language. He was also to develop Hoare Logic, enabling programmers to convert program statements into provable logical formulas. As a result, Hoare was able to establish programming as a formal science, earning his place as one of the software artists of the 20th century.

SIR CHARLES ANTONY (TONY) RICHARD HOARE

C. A. R. Hoare is a well-known computer scientist who has written and lectured broadly on the topic of the science of computing and the engineering of software. He contributed to the design and definition of many programming languages by developing a range of different algorithms and specification techniques that are used throughout the computing industry. His most important contributions include development of one of the world's most widely used sorting algorithms, called Quicksort; and his introduction of the formal logic system known as Hoare Logic.

Born in Sri Lanka in 1934, the son of British parents, he attended Oxford University in the early 1950s, where he studied the classics. He received his bachelor's degree from Oxford in classics in 1956 and stayed on for another year to study statistics. After serving two years in the Royal Navy, he embarked on a course of studies of computer translation of human languages and probability theory at Moscow State University in what was then the Soviet Union. Along the way, he became keenly interested not only in computer science in general but also in the use of logic as a method of determination of mathematical truth. In 1959, while at Moscow State University, Hoare developed the Quicksort algorithm, which was the first and one of the most efficient sorting algorithms, still in use today.

By 1960, Hoare returned to England where, until 1968, he was employed by Elliott Brothers, a small, scientific computer manufacturing company located in London. He initially worked as a programmer, but he quickly rose to the position of chief scientist. He supervised the development of the first commercial compiler for the Algol60 programming language.

One of his most important contributions to computer science was his development of Hoare Logic. During much of the 1960s and 1970s, a major issue in the field of computer science was the crisis resulting from the relentlessly increasing complexity of computer software and systems. In response to this problem, Hoare developed a set of logical rules (now known as *Hoare Logic*) that could be used by programmers to assure a solid logical foundation for their software products.

He developed Hoare Logic as a specification technique for programmers to convert programs into provable logical formulas that could be verified for correctness. He subsequently contributed to the growing field of parallel processing when he introduced the language called CSP (Communicating Sequential Processes), which was also the basis for the Occam programming language.

Between 1968 and 1977, he was a member of the faculty at the University of Belfast in North Ireland. While there, he concentrated his research on studying the differences between operating systems and compilers. He attempted to apply advances in programming language and theory to solve the problem of concurrency, the problem of how to permit access to shared resources between programs executing computations simultaneously. In 1969, he applied logic techniques now known as axiomatic semantics. This was a major leap forward in the logic development of programming languages.

Throughout his career he maintained his interest in the problem of concurrency. He also maintained a strong research interest in the unification of the theories of programming. He worked to strengthen the links between the different schools of research with the ultimate goal that the disparate theoretical approaches could be unified and used for practical applications, particularly in software engineering.

Hoare remained at the University of Belfast until 1977 when he moved to Oxford as a professor of computing. There he led the Programming Research Group of the Oxford University Computing Laboratory. He retired from that position in 1999 but continues as a professor emeritus while he also serves as a senior researcher at Microsoft Research in Cambridge, England.

As the development of the software industry progressed from its early years, a transition took place in which the systems, originally used primarily for scientific applications, began to be used for business applications. At this point, the state of software development became more attractive to industrial participation. Increasingly, individuals who had learned the skills of software development recognized the opportunity to start their own businesses and provide their services under contract.

SOFTWARE AS AN INDUSTRY

During the mid-1970s, the software industry came into its own. In part this was due to the development of more capability for cross-platform applications with the standardization of 3GLs (structured programming allowed larger-scale software systems based on existing specifications), and in part it was due to the shift into business applications; but by far the most important factor was the introduction and remarkable early success of the personal computer, which led to the founding of PC software firms such as Microsoft and Software Arts.

In 1975, Bill Gates and Paul Allen developed Basic for the Altair 8800 and created their software company, Microsoft. By 1977, Apple was enjoying great success with the Apple II, and the company Software Arts developed the first spreadsheet program, VisiCalc. This demonstrated that the personal computer could do much more than serve as a videogame machine, and it set the stage for a spectacular expansion of the software industry to meet an almost insatiable and growing demand.

By the 1980s, the methods of object-oriented programming began to have a major impact by making it easier to modify software to adapt to changes in needed functionality. Typical applications at this time were desktop publishing, spreadsheets, and other productivity-enhancing applications that were based on user-initiated mouse clicks or menu selections as well as keyboard input.

In addition to the packaged software business, the 1970s and 1980s saw the contract programming industry continue to grow at a rapid pace. These businesses provided what was known as "professional services." They often provided various services in the areas of consulting, analysis, and design, in addition to programming. Considerable effort was applied to improve productivity in software development.

SOFTWARE PRODUCTIVITY

The progress of software development from the 1940s to the 1980s was remarkable. The basic concepts of programming were developed during this thirty-year period, and programming changed from moving wires and setting mechanical switches to writing software using 1's and 0's of machine language; then to assemblers, interpreters, compilers, and operating systems; all for mainframes, minicomputers, and finally for PCs.

Software for personal computers was a revolution in itself because it democratized programming—allowing more and more natural artists to emerge.

All the while, changes in software development were being made all in the attempt to find ways to constantly improve productivity. But to improve it, it is important to be able to measure it.

One such approach to the problem of measurement that was introduced is the Halstead measure. It is based on four numbers defined in such a way as to be determined directly from a programmer's code: the numbers of interest are the number of distinct operators, the number of distinct operands, the total number of operators, and the total number of operands.

However, this and other methods of measuring productivity have not been satisfactory. Although using the number of lines of code, for example, as a measure may be appropriate within an order of magnitude, it is far from an accurate measure by itself. Another measure, the Logical Source Lines of Code (SLOC), is based on the number of executable statements. However, this at best is an inaccurate measure.

Nevertheless, while it is clear that the evolution of software generations has brought substantial improvements in the productivity of software, probably several orders of magnitude, this still doesn't compare with the nine orders of magnitude (one billion-fold increase) that occurred during the same time in the hardware domain, according to Moore's Law. Why?

One of the reasons is that, while we have been able to use microchip technology to automatically design and manufacture other microchips, we haven't yet mastered the automatic writing of even simple software. But this may be coming. One issue to decide is whether programming should be done from the top–down (command organization) or from the bottom–up (i.e., through open source self-organizing structures).

If we had good indicators of software productivity, what would they tell us? And what are the factors that influence software productivity? And how can we optimize them to improve it?

Consider a large-scale software system (LSS) that is developed by a team of developers in a top–down fashion. If productivity measures are applied to the work products of small-scale software elements, major differences, often more than an order of magnitude, can be seen for different programmers, different programs, or both; some believe that the top–down approach of large-scale programming efforts can substantially reduce this variance.

Researchers at IBM have considered this issue and have completed a characteristic study in this area.[8] They found that a major issue arises in trying to identify and measure software development productivity, cost, and size. In their effort, they characterized productivity in terms of the number of lines of code produced per person per hour.

However, it is clear that many different parameters can influence the results. These include differences in the choice of programming language used (some being more favorable with respect to the quantitative measure than others), alternative methods for program development, and the effort placed on collecting data to demonstrate measured improvement. One finding was that small project teams produced source code with more positive credit for functionality than did large teams in a comparable amount of time.

In addition, many studies don't effectively describe how they address variability in the performance of individual software writers, although it is clear that different individuals can possess markedly different capabilities. This is an indication that there are real artists in software development who are head and shoulders above the ordinary.

Consequently, to date it has not been possible to accurately quantify the level of productivity improvement that could be expected as software methods advance, and neither has it been possible to come up with truly effective measures of individual or collective contributions.

FOURTH GENERATION LANGUAGES

Since so much of programming is rewriting, porting, and converting applications, 4 GLs have been implemented to assist programmers in building applications without the direct symbolic coding of 3 GLs. Still, talented programmers are essential to developing new innovations and capabilities.

It is interesting to observe that a small number of instructions in a 4 GL will serve the same purpose as hundreds of instructions in 3 GLs such as Fortran or Cobol. Applications based on 4 GLs concentrate on the regularly performed tasks such as creating screen forms, requesting data, or making hard copies. The principal advantage of this is that a programmer can create an application in a much shorter period of time while the ultimate user can be involved earlier in the process through the use of simulation runs. The principal disadvantage of programs developed using 4 GLs is that they are much larger, needing much more disk space and a larger part of the computer's memory capacity to run.

As a result, it is fair to say that the most commonly used programming languages today are 10–30 years old and the 4 GLs have not yet achieved the dream of "programming without programmers."

[8] J. A. Darringer, et al., "LSS: A System for Production Logic Synthesis," *IBM Journal of Research and Development* 28(5): 537–545, 1984.

In considering the trends in software development, it could be expected that programming could increasingly involve a succession of calls to previously written and tested routines. Such an approach has been embedded in the concepts of *software development environment* and *computer-aided software engineering* (CASE).

Even though 4GL software has not yet been fully accepted, it is fair to say that, during the period from 1975 to 1990, considerable progress has been made in addressing the identified weaknesses of previous generations; as a result, the 4GL products had begun to be used in a considerable portion of new application software. Despite the promise, however, 4GLs can still be justifiably criticized.

For one thing, many have found that 4GL software has tended to be sloppily written and wasteful of computational resources. Another concern is the lack of compatibility with preexisting data and application software. In any case, although representing some progress, 4GLs have not been sufficiently successful to overtake the previous generations of programming languages.

Thus the current situation for software development is that a mix of languages at varying levels of abstraction are available and in use to support a broad range of different applications.

PROPRIETARY VERSUS OPEN STANDARDS

Standardization is one of the crucial issues for the future of the Information Revolution. While powerful bottom–up creative forces such as self-organization, chaos theory, and emergent behavior favor open architecture, top–down approaches are favored by businesses that desire their own proprietary standards.

The strategic use of standardization to achieve a competitive advantage has become an important factor in business planning. In addition, standardization provides a stimulus for creating a favorable situation for the deployment of new services and products in an orderly manner. Many recognize that a standard is never neutral but reflects the strengths and technologies of those who promote it. Participation in the standards process benefits businesses, not only in terms of new technology and industry dynamics, but also in developing strategic tools to protect current investments in networks and services.

It is no longer sufficient just to interconnect different networks; the end service aspects are even more important. There are also some less obvious advantages of participation in standards development. Participation in the prestandardization phase (which considers strategic issues, new architectures and interfaces, evolutionary scenarios, etc.) enables industry to identify major issues at an early stage. Consequently, during this early and influential phase of standards development, a company can modify internal policy and consider new product opportunities.

Typically, standards have come about in three ways:

1. A vendor may dominate a market and set a proprietary de facto standard (e.g., early telephony from AT&T, or the PC operating systems from Microsoft).
2. Standards organizations may establish open standards (e.g., HTML).
3. Vendors and markets may collaborate in ways that are not clearly attributable to any one organization (e.g., TCP/IP or VCR formats).

However, it appears now that technologies are developing too quickly for a single vendor to dominate, or for standards organizations to easily come to agreement. Where there is no clear dominant architecture or vendor, how will standards emerge? How will the seemingly endless list of standards and protocols be optimized? These issues require integration of standards teams across technology areas so that selections can be made from the various options to achieve optimization of globally interoperable standards.

A fundamental question is: Where will network intelligence be located? Will it be decentralized and distributed on an open Web architecture or will it be centralized through proprietary standard?

EMERGENT FIFTH GENERATION LANGUAGES (5GLs)

Since the 1980s, there has been progress toward a fifth generation of computers and software. This was an outgrowth of the longstanding AI movement.[9]

In the fifth generation, machine architecture reflects the composition of the system's programming language, and hardware and software are even more integrated than before. In fact, the trend toward machine reasoning is reflected in this approach. By integrating hardware and software, at both the operating system and the application levels, we can envision a merger of the generations of computing. This would absorb conventional software development models as subsets or special cases.

There have been several fifth generation initiatives in the United States, Japan, and Europe. So far, the outcomes of the various initiatives have failed to meet expectations. One approach that has been suggested is referred to as the *expert systems approach*. This approach envisions the use of domain-specialists' knowledge combined with processing by an *inference engine*. In this approach, a consulting session would be convened in which a user would provide data describing a particular circumstance or situation, and the computer would provide its recommendation for resolving an issue or problem, taking into account the additional information available in its knowledge base as well as a set of general purpose rules. Expert systems development

[9]R. Clarke, "A Contingency Approach to the Application Software Generations," February 24, 1991. `http://www.anu.edu.au/people/Roger.Clarke/SOS/SwareGenns.html`.

for such a concept may use logic programming such as Prolog or *expert system shells*.

However, significant difficulties have been encountered in applying expert systems. In some cases, the original knowledge base has had to be modified to make it appropriate to the problem at hand; the approaches to knowledge elicitation and knowledge engineering have required modification; and expert systems software modules have required adaptation to be able to be implemented in the desired applications, rather than as freestanding software.

As we have progressed from the first to the third generation of programming, the software developer needed to take an active role to define the solution for a problem. However, as we move to successive generations, the resulting applications have allowed the solutions to be expressed in a form increasingly convenient to the user. The fourth generation enables the human to delegate part of the solution to the computer, provided that a problem definition is provided in an appropriate form.[10]

With fifth generation technology, the developer is empowered to operate at a higher level of abstraction. The use of *knowledge-based technology* alleviates the programmer of the need to formulate an explicit definition of the problem. The problem definition is contained implicitly within the process.

Logic programming involves the art of using logic to describe knowledge. This objective can be met by outlining a problem in terms of facts and rules that can be expressed in first-order logic. A hidden theorem prover within the software can then be used to solve a particular problem. In some ways, logic programming can be seen as the essence of fifth generation computing. Building on existing logic tools, development is ongoing in new tools. Related areas, such as neural networks and parallelism, are also being investigated.

Thus finding new ways to manage, organize, and implement programming has been a difficult process. The fundamental issue confronting the industry has been whether the next generation of software should be designed and developed from the top–down (through an authoritative structured organization) or from the bottom–up (through self-organizing open source).

One inspired misadventure offers some important lessons about top–down programming—Charles Simonyi's meta-programmer concept that was applied to Microsoft's command programming management.

CHARLES SIMONYI

Born in Hungary in 1948, Charles Simonyi was a teenager before he saw his first computer. It was a Russian machine called the Ural II that filled the room and was built from thousands of vacuum tubes. Since his father was an engineering professor, Simonyi had the opportunity to work on the computer. The

[10]Ibid.

Ural II had 4K of memory, a miniscule amount by today's standards, about as much as one of the first personal computers of the 1970s. Following his experience with the Ural II, Simonyi decided to pursue his interest in computer programming.

His first programming effort was to write a compiler program for the government. Subsequently, at a trade fair in Budapest, Simonyi had the opportunity to interact with foreigners from Denmark. He presented a demonstration of his compiler to the members of a Danish computer trade delegation. He was soon contacted about a job. At the age of 16, he left his family and moved to the West.

Simonyi spent the next year writing software in Denmark while saving his money so that he could pursue an advanced education. He then enrolled as a student at the University of California at Berkeley and graduated in 1972. At this time, he was recruited to work at Xerox PARC. He continued his education by registering in a doctoral program at nearby Stanford University. His doctoral thesis was on the methods he had developed to write software. Charles Simonyi was an extremely talented programmer—a software artist.

For an artist to create a painting masterpiece, an artist must have superb imagination. To create a software masterpiece, a software artist must have superb memory.

Different people have different abilities when it comes to memory, but only a select few have unusual abilities to remember things. For example, it is unusual to find a person who can remember more than nine randomly selected numbers at a time. Nevertheless, it is said that it takes a person with such ability to be able to write really good computer programs.

To write software, it is necessary to keep track of the complex flow of data through a program; as a result, being able to remember several items at a time is important. The best programmers have a remarkable ability to remember complex things. As a result, there are some programmers who are 100 times more productive than the average programmer, simply because of the complexity of the computer code they can compose.

While working at Xerox PARC in the 1970s, Simonyi had difficulty getting assistants. When new staff were added, they were generally Ph.D.s who had their own research interests and priorities, and they were not generally receptive to taking direction from Simonyi. At one time, Simonyi proposed a research project to study programmer productivity and how to increase it. In the course of the study, his test subjects were paid to write software under his supervision. The test subjects were Stanford computer science students. The software they were contributing to was Bravo, Simonyi's proposed editor for the Xerox personal computer—the Alto. By referring to them as "research subjects" rather than programmers, Simonyi was finally able to obtain the assistance he needed at PARC.

The Bravo experiment was a success, and this word processing program was one of the first examples of software that presented on-screen document images that were identical to the printed output. Beyond the Bravo project, the study provided data for Simonyi's own Ph.D. dissertation titled *Meta-programming: A Software Production Method.*[11]

In Simonyi's doctoral research, he intended to develop a more efficient way to organize programmers in software writing. Since software development seems to always expand to fill the available time, his research dealt with how to get more work done in the limited time that was available.

Simonyi concluded from his data and experience that simply adding more programmers was not an effective approach to meet a deadline. The addition of programmers simply increased the amount of communication overhead on the project without a commensurate acceleration of the work progress. He found that the trick to improving programming productivity was to make better use of the programmers already in place. Simonyi's method was to create a meta-programmer.

The meta-programmer was the individual who served as the designer, decision maker, and communicator in a software development group. As the meta-programmer on Bravo, Simonyi himself mapped out the basic design for the software, deciding what it would look like and what would be the underlying code structure. But he did not write the actual code.

Simonyi prepared a document for his research subjects to use as basic guidance in writing their individual code. Once the overall program was designed, the meta-programmer handled communications. In general, Simonyi felt it was more important to the effectiveness of the project for decisions to be made quickly than that they be made well.

By centralizing the process of design, decision making, and communications, Simonyi felt that software could be developed more efficiently and faster. The key to the plan's success was finding obedient and yet effective programmers.

In the competitive work structure of PARC, only the elite could survive the demanding intellectual environment. In order to bring junior people into the development organization, Simonyi promoted the elite to the status of meta-programmer. Both the organization of PARC's Computing Science Laboratory (CSL) and Simonyi's meta-programmer system had hub and spoke management structures. At CSL, however, most decision making was delegated to the research group level. In Simonyi's system, only the meta-programmer had the authority to make decisions within a very rigid authoritarian structure.

While looking for a new job in 1980, Simonyi had lunch with a former PARC colleague, Bob Metcalfe, who had recently founded 3Com. Metcalfe gave Simonyi a list of people to contact for a new job opportunity. One on the list was Bill Gates.

As a result, Charles Simonyi left his job at Xerox PARC to work for Bill Gates at Microsoft. They shared a common vision of creating software. Gates

[11]C. Simonyi, *Meta-programming: A Software Production Method,* Ph.D. thesis, Stanford University, December 1976. `http://www.parc.xerox.com/publications/bw-ps-gz/csl76-7.ps.gz` (December 2001).

wanted applications to become more important to Microsoft than its operating system, and Simonyi was the programmer he picked to make that happen.

WILLIAM H. GATES III

Bill Gates was born on October 28, 1955 in Seattle, Washington and, along with his two sisters, was raised there by his father, a wealthy attorney, and his mother, a schoolteacher and a bank director. He attended private school, where he developed an early interest in software and computers, and he began programming computers at the age of 13. By 1972, he and his childhood friend Paul Allen purchased an Intel 8008 microprocessor chip and used it to build an automated car-counting machine. They formed a small company that they called Traf-O-Data.

Gates enrolled at Harvard in the fall of 1973 and began his studies in pre-law. He later claimed that he had decided to go to Harvard University to learn from people smarter than him, but that he left disappointed. When he arrived at Harvard, he didn't know what he wanted to do in life. While he registered as a pre-law student, he had little interest following his father's footsteps in law. His parents didn't have any expectations either. They only insisted he go to college. And Harvard was his best option.

While Gates was completing his first year at Harvard, his friend Paul Allen remained in Washington trying to find new business for their software company, Traf-O-Data. At the time, they had negotiated deals with municipalities in several states, as well as Canada, but their efforts were undercut by the federal government, which had decided to directly help cities analyze their traffic statistics.

With their company failing, Gates and Allen began considering what they should do next. They had talked about continuing their business interests by getting into the area of programming microprocessor computer systems, something they were well prepared for based on their Traf-O-Data experience. They were excited about the new advances in microprocessor technology. Surely, as these new gadgets became popular, there would be the chance to create a business opportunity. Perhaps they could write a Basic interpreter for one of the new systems that must be about ready to be released.

Meanwhile, Gates had begun thinking about dropping out of Harvard. In the summer of 1974, he interviewed for summer jobs at various places in the Boston area. He and Allen wound up working together that summer, having taken jobs at Honeywell Corporation.

In December 1974, Allen came across the January 1975 issue of *Popular Electronics*.[12] On the cover of the magazine was a picture of the MITS Altair 8080, a rectangular metal machine with toggle switches and lights on the front.

[12] *Popular Electronics*, January 1975. `http://www.computermuseum.20m.com/popelectronics.htm`.

The cover had the headline "World's First Microcomputer Kit to Rival Commercial Models." Allen later said, "I bought a copy, read it, and raced back to Bill's dorm to talk to him." He then said, "I told Bill, 'Well here's our opportunity to do something with Basic'."[13]

Gates agreed. It was time. The personal computer miracle they had been anticipating was about to happen. And their experience in programming the 8008 microprocessor would be just what they needed to create the right business opportunity.

Earlier, while Gates and Allen were finishing high school and then struggling to figure out their futures, another young entrepreneur, Ed Roberts, was proceeding down a parallel path. A hobbyist and gadget nut with enormous energy and the bulk to match, Ed Roberts loved tinkering with electronic hardware. At first, he began a company called Micro Instrumentation and Telemetry Systems (MITS), which he operated out of his garage in Albuquerque, New Mexico. Here, he sold mail-order model rocket equipment. In 1969, he moved MITS out of the garage, and he sunk all of his company's capital into the commercial calculator market. MITS was the first company in the United States to build calculator kits. They quickly expanded to more than 100 employees. Then the bottom fell out. In the early 1970s, Texas Instruments entered the calculator market, and MITS found it impossible to compete.

By 1974, MITS was in debt for more than a quarter of a million dollars. Roberts was counting on the next generation of chip, the 8080. It was much faster and had much more brainpower than the 8008. The new chip could offer much more capability than was needed for a calculator; surely it could support a small computer. So he decided he would offer a new small computer system kit, called the Altair, and he would sell it for $397. This was an astonishing figure, and Roberts knew it. After all, Intel's 8080 chip alone was selling for $350. But Roberts had been able to get Intel to sell him the chips in volume at $75 apiece.

Later in 1974, *Popular Electronics* undertook to write an article about Roberts, MITS, and the Altair. But before it could publish the article, it needed to see the prototype and verify that it actually worked. Roberts shipped his only working model to the *Popular Electronics* offices in New York City, but it never arrived. The world's first home computer had disappeared in transit!

Nevertheless, Roberts and his MITS engineers hurriedly put together a metal shell with switches and lights on the front as a mock-up, and they shipped it to New York. And it was the photo of this nonworking box that appeared on the magazine's cover.

The article on the Altair stated that the computer had 256 bytes of memory and 18 slots for additional memory boards that could increase its capacity to about 4096 bytes. There was no screen or keyboard. Since no one had developed a high level language for the 8080 microchip, the Altair could only be

[13]D. Ledgard, "25 Years Since the First Microcomputer—The Altair," June 1999. `http://www.colonization.biz/me/altair.htm`.

programmed in complex binary machine language. This would have to be done by painstakingly flipping the switches on the front panel.

Gates and Allen made a long-distance call to the MITS office in Albuquerque after reading the *Popular Electronics* article. They talked directly to Roberts, and Gates explained that they had developed a Basic interpreter that could be adapted for the Altair computer.

Roberts later recalled:"We had at least 50 people approach us saying they had a Basic. We just told everyone, including those guys, whoever showed up first with a working Basic had the deal." Roberts wanted a version that would work on the Altair, not one that could theoretically be adapted to do so. Furthermore, some of Robert's engineers had told him that the 8080 chip would not be able to support Basic.[14]

At Harvard, Gates and Allen immediately went to work to prove the engineers wrong. Shortly, they sent a follow-up letter to Roberts saying that they now had a Basic that worked with the 8080 Intel chip. They proposed to license MITS to sell their software with the Altair to hobbyists for royalties.

Having claimed that they had a working Basic interpreter, they had to produce one. For the next eight weeks, the two would work day and night, trying to develop a high level computer language for the 8080 chip.

Since Gates and Allen didn't have an Altair, they needed information about it. They obtained a manual on the 8080 microprocessor written by Adam Osborne, an Intel engineer. While Gates concentrated his efforts on writing code for the Basic interpreter, Allen proceeded to create a minicomputer emulator that could mimic the 8080 chip. "It wasn't a question of whether I could write the program," Gates said, "but rather a question of whether I could squeeze it into 4K and make it super fast."[15] Somehow, he made it work.

After contacting MITS again, Allen took the Basic interpreter with him to Albuquerque to demonstrate it on the Altair. When he entered the program into the Altair, it worked flawlessly the first time. MITS approved its deal with Gates and Allen that allowed them to retain the rights to their Basic interpreter. And Microsoft was born.

Now in his junior year, Gates decided to leave Harvard to apply his energies to the new company. Convinced that personal computers would find their way onto every office desktop and into every home, they began in earnest to develop software for them. This vision for personal computing has been a key factor guiding the success of Microsoft in particular and setting new directions for the software industry more generally.

Gates loved programming and would compete with his employees. In the book *Programmers at Work*,[16] Gates is quoted as saying: "In the first four years of the company, there was no Microsoft program that I wasn't involved in actually writing and designing." He acknowledged that his style would cause

[14]S. Segaller, *Nerds 2.0.1: A Brief History of the Internet*, TV Books, New York, 1998.
[15]Ibid.
[16]S. Lammers (ed.), *Programmers at Work*, Microsoft Press, Redmond, WA, June 26, 1986.

some friction with his programmers. "It's kind of painful sometimes if you have somebody else working on the project. They never code stuff exactly the same way you like to see it coded. I remember when we were working on Basic, I'd go back and recode other people's section of code, without making any dramatic improvements. That bothers people when you go in and do that, but sometimes you just feel like you have to do it."[17]

Following the early flurry of excitement with the Altair, the IBM PC soon eclipsed the Apple II and every other machine on the market. By the end of 1983, IBM had sold more than a half million PCs. At the time, there were four types of software in the microcomputer business: operating systems like Gary Kildall's CP/M, programming languages like Bill Gate's Basic, applications like VisiCalc, and utilities. Gates knew very little about applications, but when Charles Simonyi joined Microsoft in 1979, he brought with him his experience in developing application software.

Simonyi came straight from PARC and brought with him exactly the expertise that Gates needed to start an applications division at Microsoft. They quickly made a wish list of products to develop, including a spreadsheet program, a word processor, and a database package. Simonyi was one of many PARC transplant successes.

Simonyi also brought his Ph.D. research to Microsoft. Reading through the dissertation, Gates saw in Simonyi's meta-programmer concept just the mechanism he thought he needed to control Microsoft and its fifty employees.

The term meta-programmer was not used at Microsoft. Gates called his system at Microsoft a "software factory," but what he and Simonyi implemented at Microsoft was a hierarchy of meta-programmers. Unlike Simonyi's original vision, Gates implemented several levels of meta-programmers, which allowed a much larger organization. Gates was the lead meta-programmer.

But after less than three months, the meta-programming experiment was a failure according to at least one observer.[18] The lesson was that art cannot flourish in an autocracy. This highly specialized top–down approach proved inefficient and counterproductive.

Software development, like the writing of books, is an iterative process. You write a program and if it doesn't work, you improve it. The information flow was never adequate to demonstrate success. Microsoft went back to writing code in the conventional manner. But the structure of architects and program managers was left in place.

Microsoft had managed to keep its development teams small, even as the company had grown. In designing, writing, and integrating operating system

[17]J. Wallace, and J. Erickson, *Hard Drive: Bill Gates and the Making of the Microsoft Empire*, Harper-Collins, New York, 1993.

[18]R. Cringely, *Acidental Empires: How the Boys of Silicon Valley Make Their Millions, Battle Foreign Competition, and Still Can't Get a Date*, Reprint edition, HarperCollins, New York, 1996.

software with application software so successfully, Microsoft has dominated the software development process for decades.

Has this concentration of standardization and uniformity been healthy for software innovation? Don't the concepts of self-organization, chaos, and emergence favor an open architecture and environment? In contrast to Microsoft's authoritative programming structure, consider open source development.

Let's look at the contributions of Linus Torvalds, an innovator who is a leading promoter of the open standards approach. Linus Torvalds is the brilliant mind behind Linux.

LINUS TORVALDS

Linus Torvalds was born in Helsinki, Finland in 1969, the son of journalists and the grandson of a Finnish poet. He pursued university studies at the University of Helsinki from 1988 to 1996, and received his master's degree in computer science in 1996. His master's thesis was entitled *Linux: A Portable Operating System*.[19]

Torvald's interest in personal computing started when he was a child. By the time he was about 20 years old, he expanded on that interest and embarked on his efforts to create the Linux operating system, stimulated by his desire to see if he could improve on the UNIX operating system, an open-standard operating system that has been in widespread use since its development in the 1960s.

Linux is an important example of open source development and user generated software; its underlying source code can be used, modified, and redistributed by anyone freely. In constructing Linux, Torvald posted his early versions on the Web and solicited input from computer science experts throughout the world to enhance it. He was not the first person to use this approach to improve open source software, but he may well have been the most successful.

Not only is Linux an excellent example of the power of the open source approach, it should be pointed out that it has also proved to be an extremely stable and effective operating system, with particular value when used in the role of a network server. As an open source system, it is still constantly improving and evolving. At the present time, as much as 80% of the servers of the most reliable network hosting services now use Linux as their operating system.

Linux, like UNIX, is a major democratic operating system and social force in computing. One of the main achievements of Linux is its counterbalancing influence on Microsoft—in effect, positioning itself at the other end of the open versus proprietary standards issue.

[19]L. Torvalds, *Linux: A Portable Operating System*, Master's thesis, University of Helsinki, 1996.

PATTERNS OF DISCOVERY

Software artists like von Neumann, Shannon, Hoare, Simonyi, Gates, and Torvalds were trying to create efficient and easy-to-use programming languages to logically process information and data. They had to trade off the level of abstraction against computational efficiency. Through their efforts, they invented key programming concepts, methods, and applications. As a result, each generation of programming language worked to overcome the limitations of its predecessor.

The von Neumann, Shannon, and Hoare discoveries followed Proof of Principle Patterns. However, programming efforts, including those of Simonyi, Torvalds, and Gates, primarily followed the 1% Inspiration and 99% Perspiration Pattern.

FORECAST FOR CONNECTING PROCESSES

One of software's perplexing issues that succeeding generations of software have not been able to escape is that Moore's Law changed hardware by a factor of a billion, but no such productivity gains have been seen for software as yet. Why is that the case?

The meta-programmer experiment at Microsoft showed that top–down authoritative programming crushes creativity and productivity. Perhaps we have missed a paradigm shift. Somewhere serendipity should have opened the door to a new way of programming and it was either missed or will occur very soon. One possibility is software development returning to the bottom–up approach, once again releasing the talent of the artist.

Discoveries Requiring Inspiration and Perspiration

The software industry can be expected to continue to exploit innovation to produce a variety of improvements, including portable translation devices for simple conversation in any language; natural language processing for speech recognition, control, and operation of all devices; improved 3D collaboration tools for enterprises such as groupware and product lifecycle management; design and animation tools with improved security software; secure digital identity; and micropayments to facilitate Web commerce.

Discoveries Requiring New Proof of Principle

Over the next twenty years, several areas appear ready to exploit a new principle in the areas of virtual reality, software agents, and adaptive algorithms.

Virtual reality can be expected to attract the interest of greater numbers of people in their recreation, competition, socialization, entertainment, and business transactions. We can anticipate the introduction of life-size, 3D images and connections to the human nervous system. Virtual reality may come to mean more to some people than actual reality.

And software agents are going to become more important. A software agent is software that acts for a user with the authority to decide when (and if) action is appropriate. Intelligent agents will use AI for learning and reasoning.

AI research can be expected to grow dramatically. The design of intelligent software is an area of AI research that is receiving great attention. Many companies and universities have recognized this trend and are engaged is serious research in this area.

In addition, adaptive software requires sensing the environment and reconfiguring in response. Adaptation implies learning, and this is another priority area of research. The next generation of software will include adaptive software.

Next generation software can be expected to do more for us because of our increasingly complex environments. The complexity comes from users, systems, devices, and goals. Programmers were accustomed to trading off CPU time against RAM space; now they must also worry about bandwidth, security, quality of information, resolution of images, and other factors.

Adaptive software offers to change this by adding a feedback loop that provides information based on performance. The design criteria themselves become a part of the program and the program reconfigures itself as the environment changes.

Concepts borrowed from biology, such as evolution by natural selection, are being introduced into AI to solve complex and highly nonlinear optimization problems. One such tool is genetic algorithms, a form of machine learning that mimics the process of biological evolution. Individual software units are referred to as chromosomes and consist of genes or parameters of the problem being optimized. The "fitter" program elements are chosen to reproduce while the remainder is eliminated in succeeding generations. After a number of generations, the algorithm should converge on the chromosomes representing an optimal solution.[20] It is worth noting the implicit parallelism of genetic algorithms.

While genetic algorithms exist mostly as research activities at academic institutions and commercial applications are still largely in developmental stages, they do offer Web applications the potential ability to adapt to their environment.

In the digital world, genetic algorithms, capable of adapting to their environment faster than their competition, can obtain a significant advantage for survival. Already software engineers are introducing genetic algorithms for the Web, but progress in this area would most likely benefit from the development of a language with specific qualifications for these types of applications.

[20] M. Lacy, "An Introduction to Genetic Algorithms in Java." `http://www2.sys-con.com/ITSG/virtualcd/Java/archives/0601/lacy/index.html`.

Discoveries Requiring New Serendipity Within Twenty Years

Just as Moore's Law captures the future of hardware technology capacity, a new "law of software productivity" may emerge to enable ubiquitous intelligence. One possibility is through bottom–up emergent self-organizing systems.

Self-organizing systems can take the form of collections of individual entities whose individual behaviors result in a higher level of collective behavior without a higher level of intelligence to guide that behavior. A good analogy of this is the behavior of ant colonies. The colony's collective and purposeful behavior is an emergent behavior that results from the seemingly nondirected, specialized behavior of the individual ants. In such colonies, certain members perform tasks without direction, such as foraging for food, cleaning up waste material, carrying out the dead, and fighting off invaders. Colonies modify their behavior and learn over time, although it is not believed that individual ants are capable of such learning.

Nevertheless, the emergence of new capabilities in software will likely require some human direction, at least initially. Emergent software would be intended to solve problems for which we do not have an *a priori* solution approach. Emergent software doesn't need to be exotic: game designers use the principles of emergent behavior to provide players with ever-changing game experiences—just take the SimCity software as an example.

Evolutionary algorithms have several key questions: How long would such an evolutionary system need to run before it begins to exhibit a distinct observable change in its organizational phase? How would we recognize such a change in phase within an autonomous, self-organized computing network? What is the right level of complexity?

Computers are able to handle levels of software complexity that exceed what is needed to simulate parallel, self-organizing, and fractal algorithms that mimic the human brain. In addition, there have been dramatic improvements in the speed of software algorithms operating on the same hardware. These improvements vary for various problems, but are pervasive.

5

Connecting Machines

The easier it is to communicate, the faster change happens.
—James Burke[1]

The United Nations (UN) is an international organization created in 1945 to foster global peace and security, friendly relations among nations, and international cooperation in addressing economic, social, cultural, or humanitarian issues. The UN represents an important connection among the nations of the world. Headquartered in New York City, the member countries, presently numbering 192, deliberate on issues of international importance and set policies and guidance to improve the human condition and avoid conflict. Protocols and language conventions are an integral part of the operations at the UN and are used to promote open and orderly communications among the diverse membership.

In a similar fashion, the multitudes of computer systems that are connected through networks must adhere to protocols and language conventions to enable effective communications and facilitate network harmony.

Networking has had a major impact on human society and culture by *connecting machines*, particularly with the advent of the Internet and the World Wide Web. By linking machines on a global scale, the power of connections has found new ways of enhancing communications, business processes, and entertainment while bringing dramatic improvement in productivity and opening new channels of communications.

Ethernet technology has been the key in enabling machines to effectively interconnect, and individuals such as Robert Metcalf have been central players in bringing this about. In this chapter, we present the Ethernet story, with a

[1]J. Burke, *Connections*, Revised edition, Little Brown & Co., Boston, 1995.

Connections: Patterns of Discovery By H. Peter Alesso and Craig F. Smith
Copyright © 2008 John Wiley & Sons, Inc.

description of the role of Xerox Palo Alto Research Center (PARC), Robert Metcalf, David Boggs, and the events that led to our present ability to connect machines into a global network.

THE ETHERNET STORY

Initially, computers were stand-alone machines. They were connected only to their immediate peripherals such as printers. In order to share information or transfer information from one machine to another, it was necessary to record the data in some transportable form such as punched cards or magnetic tape, physically carry this transfer medium to another compatible system, and then go through the process of reading the data into the second computer. Networking technology made it possible to connect computing machines to each other and to their supporting peripherals for the exchange of vital data and sharing of communal resources. As corporations became increasingly dependent on their computing resources, and as these resources became more decentralized, the need for networking interconnections became more and more important. Meanwhile, corporate mergers, takeovers, and downsizing provided additional impetus to increase the sharing of corporate data in ways that were fast and seamless while supporting multiuser access to centralized data stores or integrated databases.

In addition, companies began to see that their data could be effectively moved off expensive mainframe computers and onto smaller machines. Also, these machines could be economically interconnected with local area networks, and executives began to direct their internal Information Technology (IT) organizations to redesign their corporate systems to facilitate the transfer of applications and databases. With these driving forces, corporate local area networks began to increase in number and size, and system designers began to develop new ways to interconnect not only computers but also networks themselves.

Speed has always been the controlling factor that limits the capabilities of networks. At the present time, speed remains an issue but we have seen great improvement in a relatively short period. Network architectures and protocols have become increasingly complex, as networks have been extended locally into individual offices, small businesses, and private homes, while at the same time expanding to global reach. Today, it is usual for an email message or a piece of corporate data to travel to recipients across the globe almost instantly. This is a testament to the speed of our networks today, yet the rapidly growing volume of data represented by voice communications, graphics, and video is increasingly adding to the traffic.

Today's vast intertwine of networks includes several different types of transmission media including fiberoptic cable, twisted pair (copper wire) cable, coaxial cable, microwave radio links, and infrared connections such as Bluetooth. The two dominant media in current use for large-scale data transmissions are fiber optics and unshielded twisted pair cabling. Fiber optics provides the higher speed and larger capacity because it is a nonelectric medium. In contrast, not only is copper wire slower, it also can act as an antenna and therefore be subject to noise and interference. Nevertheless, copper wire is a major factor in today's network systems if for no other reasons than its cost advantage and the fact that there is a large investment in existing infrastructures that have a great reliance on twisted pair cabling.

This immense interlacing structure of fiber and wire is organized, interfaced, and coordinated by different levels of networking. Differentiating the network types is based on bandwidth and physical extent: local area networks (LANs) typically interconnect computers that are located within a small distance, usually up to 5–10 km, such as within a building or on a university or corporate campus, through a set of connections that operate at a moderate bandwidth. Metropolitan area networks (MANs) connect several LANs and usually cover a larger geographical area of up to 10–100 km in distance. Wide area networks (WANs) interconnect networked computer systems at distances of 100–1000 km through broad bandwidth connections. Finally, global area networks (GANs) connect networks between countries and across continents and oceans to provide global reach.

Several different operating systems, including UNIX, Linux, and Windows, are used on the servers that administer the networks.

The term *Intranet* refers to networks that connect computing resources within a school or corporation, but in contrast to the Internet, access and users are limited, monitored, and controlled. On a broadcast type network, such as is used in the Ethernet technology, any system that is connected to the cable can transmit a message. In such systems, when messages collide they can become garbled and problems can result. The extra obligations on the system to manage the problem of collisions results in a reduction in the transmission rate.

Ethernet was one of the original methods invented specifically to enable the interconnection of computers. In its original form, Ethernet was designed to operate over shared coaxial cables that would provide the medium for broadcast transmission; broadcast transmission occurs when information is sent from one point on a network to all other points. Because the cable used to carry communications was analogous to the "ether," which was said to be the medium for radio transmission, the name Ethernet was selected. The original patent for *Ethernet* described the essence of the technology as a "multipoint data communication system with collision detection." [2]

Ethernet technology has matured over the years. The original Ethernet design called for a transmission rate of 3 megabits per second (Mbps), which

[2] "The Ethernet Effect: Collaboration, Interoperability and Adoption of New Technologies," A University of New Hampshire InterOperability Laboratory White Paper in Collaboration with Dell'Oro Group, April 2006. `http://www.iol.unh.edu/services/testing/fe/training/The_Ethernet_Effect_WhitePaper.pdf`.

at the time was considered to be lightning fast. The standard Ethernet speed has since increased to 10Mbps. More recent higher speed Ethernets include the Fast Ethernet, the switched Ethernet, and the Gigabit Ethernet. The Fast Ethernet is a system based on a shared protocol that reaches speeds of 100Mbps, ten times the standard Ethernet used by most LANs. Switched Ethernet is a nonshared service; devices within the network are given their own dedicated paths within a LAN and, as a result, network congestion is reduced. Gigabit Ethernet works with existing LAN protocols but offers transmission rates of up to 1000Mbps through the use of fiberoptic cabling.

LANs normally can be connected to WANs through a gateway, which is a computer or dedicated device that has multiple network connections. It provides the function of converting data traffic to make sure that it has the proper format for the network into which it is being sent.

Routers are another type of networking device that play an important role by communicating with each other and maintaining and sharing information about which transmission routes are available and providing directions for messages to reach destinations. Routers also are used to connect between LANs and WANs, while they also translate protocols and determine the best path for data traffic to take to reach a destination.

When it comes to networks, it is reasonable to ask: How fast is fast enough? The desired speed always seems to be just a little faster than what is currently available. We have seen a natural evolution of network technology in which an ever increasing bandwidth is needed to deliver increasing amounts of data over the network, which is comprised of a variety of different links, each with their own transmission limits—from the fiberoptic backbone of the Internet to the "last mile" lower speed connections into homes, offices, and small businesses.

The Internet has recently been upgraded to accommodate the increasing demand. Transfer rates of up to several gigabits per second already exist on some networks and will soon be more widely available. As with many of the innovations that enabled today's computer revolution, the beginning of much of our networking technology can be traced back to the efforts at Xerox PARC in the1970s.

XEROX PARC AND ETHERNET

Xerox PARC was the center of development not only for the world's first personal computer—the Alto—but also for many of the ancillary concepts of computing including the graphical user interface, the high-resolution display, the word processor, the mouse, and the technology for connecting computers—networking called Ethernet.

Today, Ethernet is the dominant networking technology. It provides the backbone of corporate networks and is an important feature in wireless WiFi networks as well. The idea for Ethernet was the result of the work of a brilliant inventor, Robert Metcalf, whose technology breakthrough was based on the need to address organizational requirements.

The concept of time sharing had been investigated from the 1950s as a means to allow multiple users to simultaneously use centralized computing resources. By connecting multiple users at remote terminals to the main computer, the users could concurrently access the computer as long as it was designed to quickly and invisibly switch between the activities of the different users. In time sharing, the computer's operating system allows each user to access the computer capabilities as if he were the sole user. Of course, the remote terminal, which could be considered to be a computer device with limited capabilities, needed to be connected to the main computer. That's where the networking comes in.

In spite of the rapid progress made in the connection of remote terminals to mainframe computers, by the time 1970 rolled around, the idea of connecting computers through networks was still new. In fact, one of the most important precursors to the modern Internet, the ARPANET, had just come online in 1969, and networking was just at the starting gate for its impending sprint into the modern computing era.

In 1970, the Alto project at Xerox PARC was just getting under way. At first, the organizers of the project found they had a gap in their capabilities in that they didn't have sufficient staff with expertise in the area of computer connectivity. Although PARC had successfully connected its two minicomputers readily enough, this was a far cry from being able to connect an organization's personal computers directly to a network such as the ARPANET. It was clear that, for the desired application of the Alto—a personal computer that could form the basis for a company's distributed computing and office productivity needs—they needed to be able to connect their smaller personal computers with the larger multiuser minicomputer while concurrently obtaining access to print and file-sharing capabilities.

Although PARC management found itself without the right expert to pursue this aspect of its overall technology vision, management became aware of a graduate student who was working there by the name of Charles Simonyi. Simonyi developed a networking concept called SIGNet, short for Simonyi's Infinitely Glorious Network. This approach turned out to be unworkable and, although Simonyi went on to make significant contributions at PARC and eventually Microsoft Corporation, PARC management once again needed a lead expert for this critical part of their overall concept.

They turned to Robert Metcalf.

ROBERT METCALF

Robert Metcalf was born in 1946 in Brooklyn, New York. After graduating from high school in 1964, he enrolled at MIT, where he undertook studies in both electrical engineering and industrial management. He graduated from

MIT in 1969 with degrees in both majors and continued on to graduate school at Harvard University, studying applied mathematics. As a graduate student, he had the opportunity to work on the effort to connect to MIT to ARPANET. Subsequently, he prepared a pamphlet entitled *Scenarios for the ARPANET*, which described 19 scenarios for using ARPANET.[3] Having earned his master's degree in 1970, he continued on with his Ph.D. studies.

Here he ran into a stumbling block. Since he had devoted much of his effort working on the hardware to connect MIT to ARPANET, he focused his dissertation research on this topic. When he completed his dissertation research on these efforts, it was rejected by Harvard because it wasn't sufficiently theoretical. Fortunately for him, his persistence in pursuing computer networking paid off; he eventually received his Ph.D. from Harvard for his follow-on network development work after joining PARC. Not that this experience with academia was a happy one. Metcalf later recalled, "They let me go into this thing and they gunned me. I'm even willing to stipulate that it wasn't very good. But I'd still justify my anger at those bastards for letting me fail. Had they been doing better jobs as professors, they would never have allowed that to happen. But I hated Harvard and Harvard hated me. It was a class thing from the start." [4]

In the meantime, his experience in networking was exactly what PARC needed, so they recruited and hired him as the networking expert they so desperately needed. PARC had already developed the world's first laser printer, a speedy one-page-per-second, 500-dot-per-inch device, and they realized that they needed Metcalf's network expertise to be able to connect the Alto to its laser printer. The printer was so fast that the problem they saw at the time was how to keep the printer busy. When Metcalfe was hired, PARC also brought on David Boggs, an electrical engineer from Stanford University. The two formed a team to work on developing the technology for local networking.

Early on, Metcalf had become convinced of the promise of packet switching (as opposed to circuit switching) in communications, including communications among computers on a network. In fact, when he began to formulate his concepts for networking technology, Metcalf was strongly influenced by the packet-switching wireless network developed at the University of Hawaii–Manoa to connect computers on several different islands. This network was known as the ALOHAnet.

The design of ALOHAnet used a packet-switching method that allowed computers to transmit whenever they had data. They would consider the transmission to be complete when they received confirmation from the destination computer that all the packets had been successfully received. If, due to packet collision or any other factor, confirmation was not received, the sending computer would retransmit the data after waiting a short, randomly selected period.

[3]S. Kirsner, "The Legend of Bob Metcalfe," *Wired* 6(11): November 1998.
[4]Ibid.

Metcalfe recognized several problems with the ALOHAnet approach, so he developed some recommended modifications and made this the topic of his new dissertation. This topic resulted in the successful completion of his Ph.D. at Harvard. The main modification he offered was that if the interval for retransmission were varied based on the density of traffic, the efficiency of transmission could be significantly improved.

As Metcalf and Boggs began to scope out their effort to develop networking technology for the Alto, they realized that the network would need to be fast enough to keep up with the laser printer and would also need to have the ability to connect to other computers, perhaps as many as hundreds of them.

Within a short period, in 1973 the two young engineers developed their concept for local area networking and they called it Ethernet. Not surprisingly, it was based in part on the ALOHAnet technology. In a sense, the new technology was a modified version of the ALOHAnet system, where cables were to be used instead of radio waves to transmit the data.

The initial Ethernet was capable of sending 2.94 Mbps of data via coax cable between machines separated by distances of up to 1 kilometer. One of the key features they included in Ethernet is something called *carrier sense*; this feature works by instructing individual computers on the network to listen before they transmit data packets, thereby improving the chance that data would be successfully transmitted on the first try without a collision. In 1976, Metcalf and Boggs prepared a paper describing their concept entitled "Ethernet: Distributed Packet Switching for Local Computer Networks." [5]

The Ethernet concept worked out quite well, and its success can be seen in retrospect by the dominance that it has established as the world's standard for networking. Following its development, Metcalf began promoting the concept with the same enthusiasm he put into developing it. In 1979, he founded a new company called the 3Com Corporation (short for Computers Communications Compatibility) and through this very successful venture continued to promote the adoption of Ethernet as the networking standard.

During the 1980s, local area networking became increasingly popular, stimulated in great part by its use at many universities, where terminals and computer workstations were interconnected using Ethernet. LANs were also beginning to be adopted for business applications as is frequently the case when technologies migrate outward from their use in universities. When the Internet blasted onto the scene, it was not difficult to integrate these LANs into the global system, thereby enabling ready institution-to-institution and corporation-to-corporation communications. And so, the introduction of Ethernet was an important contributing factor to the remarkable success of the Internet.

After its introduction and widespread dissemination in LANs, Ethernet became an open standard under the Institute of Electrical and Electronics

[5]R. M. Metcalf and D. R. Boggs, "Ethernet: Distributed Packet Switching for Local Computer Networks," *Communications of the ACM* 19(7): 395–404 July 1976.

Engineers (IEEE). "Intel wanted to sell chips, Xerox printers and DEC minicomputers. None of us planned to make money selling the network, so making it an open standard was not a big risk," Metcalfe later remarked.[6]

Meanwhile, Metcalfe's company 3Com put forward the concept of using standards such as Ethernet and TCP/IP (Transmission Control Protocol and Internet Protocol) to promote connectivity. Metcalfe said, "Ethernet's most important legacy is the Internet itself. Ethernet is the on-ramp for the Internet. TCP/IP and Ethernet were both invented in the same year, and they both grew up together." [7]

Ethernet has demonstrated the possibility of commercial success with open standards. Metcalf convinced Intel Corporation to make chips for networks and he promoted the formation of the DIX (Digital-Intel-Xerox) consortium to carry the open standards banner to the IEEE's standards committee meetings. His success led to the development in 1980 of a joint proposal for a 10-Mbps Ethernet specification by the consortium.

The PC's explosive growth during the 1980s was facilitated by the prominence and availability of Ethernet. During this period, not surprisingly, 3Com experienced great financial success with soaring annual sales in hardware to support networking.

However, that is not the end of the Ethernet story. When the 1990s dawned, the blazing speed of Ethernet, 10 Mbps, no longer seemed so fast. With the rapid growth of personal computing, everything was becoming faster and cheaper, yet Ethernet appeared to be in a static mode. Realizing this situation, a group at 3Com came together to brainstorm options, in particular, options for improved Ethernet-based home automation systems. The gauntlet was thrown down when one of the group suggested, "Why can't we just make Ethernet run ten times faster?"[8] Since the key obstacle to faster networking was the same problem encountered when Ethernet was first developed—the problem of collisions between packets—and since it was obvious that simply increasing the speed would generate an unmanageable level of collisions, this did not seem to be a workable approach. Nevertheless, another engineer in the group suggested a different approach—the use of switching in networks to eliminate collisions. That group went on to form a new company, called Grand Junction, that successfully built what is now known as *Fast Ethernet*, a technology that dramatically increased the information flow rates to 100 Mbps using conventional media such as ordinary telephone wire.

Despite serious competition from other networking technology options, Ethernet has remained dominant in connecting devices. Following the introduction of Fast Ethernet, a new approach that capitalizes on the superior qualities of fiber optics has been introduced. Such technology is capable of supporting transmission speeds of 1000 Mbps.

[6]T. Mayor, "Inventing the Enterprise," *CIO Magazine* December 15, 1999–January 1, 2000 issue.
[7]Ibid.
[8]"Out of the Ether," *The Economist*, September 4, 2003.

PATTERNS OF DISCOVERY

Robert Metcalf, along with David Boggs, was trying to connect machines, overcome bandwidth restrictions, and avoid data collision during transmission. The result was the Ethernet. Ethernet has been called the on-ramp for the Internet. Metcalf and Boggs published the defining paper on this process entitled "Ethernet: Distributed Packet Switching for Local Computer Networks."[9]

Today, Ethernet is the backbone of wired and wireless corporate networks as well as the on-ramp for the Internet. The pattern of discovery was a Proof of Principle Pattern that began from a brilliant inventor's breakthrough derived from organizational principles.

FORECAST FOR CONNECTING MACHINES

We are now on the verge of seeing new and unique applications of networking. In one suggestion, all the electrical devices in a building would be controlled by an Ethernet connection; the ability to achieve this must be attributed to a great extent to Xerox PARC researchers, who not only developed the initial Ethernet technology but also had the foresight to build in a 48-bit addressing scheme, sufficiently large to accomplish this mission and much more. In fact, this vast addressing capability may well come into play as we begin to realize that some 98% of the world's microprocessors are not in what we traditionally call computers at all, but rather they are embedded in devices, appliances, and equipment scattered around the world, but not connected, at least for now, to any network.

With the continuing importance of networking technology and the need to provide for rapid growth and dramatically increasing data flow requirements, we can expect the networking industry to continue to innovate and develop new concepts.

Discoveries Requiring Inspiration and Perspiration

Inspired innovations in wireless networks and grid computing are a natural progression from current developments.

We can expect that major international communications and internet providers will support the expansion of wireless LANs, making broadband services increasingly available. Wireless protocols and technologies now support wireless data transfer rates of 54 Mbps with ranges of tens of meters. The next anticipated advance in wireless protocols and technologies is expected to increase this to 155 Mbps, enabling the interconnection of wireless LANs to create MANs over distances of ~50 km.

[9]R. M. Metcalf and D. R. Boggs, "Ethernet: Distributed Packet Switching for Local Computer Networks," *Communications of the ACM* 19(7): 395–404 July 1976.

In addition, tiny devices called motes (short for wireless transceiver and remote sensor) can be expected to create wireless sensor networks. Such networks of sensors and transceivers will enable enhancement in the awareness of movement of people and objects, and the sharing of data to allow self-organization to take place. Applications may include traffic, buildings, and ecosystems monitoring.

Ultimately, wireless technology can be expected to expand to provide continuous coverage to mobile devices. This goal of ubiquitous computing will make much of the entire world a "hot spot."

Grid computing, the sharing of computing resources of decentralized processors coordinated through networking, is another promising area. At the present time, a computer can process billions of instructions per second, but the powerful capabilities of an idle computer are just lost. Computer grids would act like a utility grid, providing greatly enhanced utilization of resources while providing increased capabilities in the hands of millions.

An example of grid computing is SETI (Search for Extraterrestrial Intelligence). It has successfully used distributed computing to analyze data related to the quest to discover intelligent life in the universe. The SETI@home project harnesses the power of the Internet to analyze signals received by radio telescopes. SETI@home distributes data to volunteers' computers. Using small software applications developed by SETI, these volunteers analyze anomalies and send the results back to SETI.

Discoveries Requiring New Proof of Principle

New proofs of principle can be expected to be developed in the areas of self-aware computers, personal health monitoring, and networked power.

People are aware of their own conditions and act to correct problems. Likewise, specially designed autonomic computers can be expected to have the ability to predict failures, repair software, anticipate user actions, prioritize workload requirements, and learn from experience. Initial elements of these behaviors have already appeared in software. Adaptive software and intelligent bots are forerunners of this technology.

Another interesting innovation area will be in health care. Virtual house calls will be possible as the elderly population increasingly need quality health monitoring in their own homes. The Internet would connect the doctor to the patient, who could then prescribe the necessary treatment, drugs, and therapy.

Innovations in the area of electric power distribution can lead to significant improvement in energy delivery efficiency. Electricity is delivered through a national power grid and energy upheavals have occurred due to the distributed generation of electricity. Monitoring and software analysis capabilities will need to be expanded.

6

Connecting Networks

I am not young enough to know everything.
—Oscar Wilde[1]

It was the Greek civilization that first conceived of an archive as a repository of data. It took them hundreds of years to find and record enough information to warrant the construction of the earliest libraries. But by the third century BC, the ruler of Egypt, Ptolemy II, became the foremost patron of knowledge when he founded the great library of Alexandria, which contained over five million books.

Today the entire concept of collecting and storing information has been influenced by the Internet and the World Wide Web. Hardly anyone has been left untouched, and even with the "technology divide" between the developed and developing parts of the world, the impact has been profound. As recently as the 1980s, few outside the academic world were even aware of the existence, much less the potential, of a global information "network of networks."

Specifically, over the past forty years, the Internet has grown from a Cold War concept for responding to the unthinkable effects of a nuclear war, into the Information Superhighway. It started from a remarkable convergence of a few key ideas related to sharing repositories of data, and the concept of robustly *connecting networks*, communications, and computation capabilities as a defense against the threat of nuclear war. Suddenly scientists and academics became aware of the potential for reaching people and computers in a new way. ARPANET, precursor to the Internet, was born.

In short order, a new burst of technology expansion was stimulated by the introduction of enhanced protocols for multimedia content, data transfer, and

[1]O. Wilde, "Quote DB." `http://www.quotedb.com/quotes/111`.

Connections: Patterns of Discovery By H. Peter Alesso and Craig F. Smith
Copyright © 2008 John Wiley & Sons, Inc.

communications, and the World Wide Web followed. It didn't take long for ordinary people in all walks of life to participate in this explosive burst of accessible technology.

Just as the railroads of the 19th century connected raw material to factories and markets in the Industrial Revolution, the World Wide Web now connects data on a global scale in the Information Revolution. In this chapter, we tell the story of the development of the Internet and the World Wide Web.

THE INTERNET STORY

The end of World War II introduced a period of carefree optimism in the United States as life returned to relative normalcy for many individuals and families. Veterans returned from the war to be educated under the GI Bill, and many new families were formed as the foundation was set for the postwar baby boom. In spite of the threat from the Cold War confrontation with the Soviet Union—a threat that was highlighted by the beginning of the nuclear arms race and the Korean conflict at the beginning of the 1950s, there was a sense that economic prosperity would continue.

On October 4, 1957, the sense of postwar tranquility and complacency in the United States was shattered when the Soviet Union launched the first earth-orbiting satellite, *Sputnik 1*. This event captured the world's attention while catching American society off guard; it created an instantaneous and widespread concern that a technology gap had developed, and it mobilized sentiment in the United States that major changes would be needed to enhance the technological competitiveness of the United States.

Among the changes that resulted from this pivotal historic event were the passing of the National Defense Education Act[2] to improve science and mathematics education, the formation of the National Aeronautics and Space Agency (NASA), and the creation of the Advanced Research Projects Agency (ARPA). ARPA was given the charter to complete "performance of such advanced projects in the field of research and development as the Secretary of Defense shall, from time to time, designate."[3] Fundamentally, ARPA had the mission to keep U.S. defense technology ahead of its adversaries, mainly the Soviet Union.

While the launching of the *Sputnik* satellite brought great attention to the need to improve education and technological readiness of the United States, it had an even more important effect in terms of highlighting the superpower struggle between the United States and the Soviet Union.

With the nuclear arms race between the superpowers, the need to plan for and defend against the possibility of nuclear war became a major consider-

[2]U.S. Congress, "National Defense Education Act," *Public Law 85-864*, 1958.

[3]U.S. Department of Defense, DoD directive 5105.15 establishing the Advanced Research Projects Agency (ARPA), signed on February 7, 1958.

ation. In the early 1960s, the Cuban missile crisis heightened fears of the nuclear threat and government agencies sought ways to contribute to the defense against nuclear catastrophe.

The basic idea was to radically change the way people, computers, and networks were connected. Originally the idea developed from a series of Air Force sponsored studies[4] by the Rand Corporation, led by Paul Baran, to consider how to protect the nation's communications and computations infrastructure in the event of a nuclear attack. The concept envisioned networks that would allow the military to maintain the command and control of nuclear arms even if major infrastructures were destroyed. The studies evaluated centralized, decentralized, and distributed systems and found that a distributed network structure offered the best survivability.

A system that provided great redundancy, reconfigurable interconnections and a new form of digital communications was the key. The system would be a network of unmanned computer nodes that could act as communications switches, routing information from one node to another until they reached their final destinations. High information flow could be achieved by breaking up a data set into multiple packets and sending the packets through the network by different paths until they all hit the final destination. In the case of a nuclear attack, packets of information could continue to find their way through the surviving portion of the network. A key element was the idea of dividing information into packets and marking their origin and destination.

Several people worked in parallel during the 1960s to develop the foundation for this new communications approach. In 1961, Leonard Kleinrock of MIT and later UCLA published a paper on packet switching theory, "Information Flow in Large Communication Nets".[5] Rand's Paul Baran suggested packet switching as part of the solution to the problem of achieving a robust, survivable computer network, and he followed up on this in his 1964 paper on data communications networks, where he proposed a communications concept involving packet-switching networks with the means of avoiding single outage points. In 1965, the British mathematician Donald Davies of the National Physical Laboratory also explored the idea of networks sending pieces of data in units from point to point across a network and he introduced the term *packet*. Other researchers also contributed to the fast-growing areas of research that included topics such as network and queuing theory.

By 1962, the recently formed ARPA had begun to focus its attention on information processing technology and formed the Information Processing Technology Office (IPTO) with Dr. J. C. R. Licklider heading the effort. In particular, ARPA had become interested in networks of computers that could share resources, the technology known as time sharing. More broadly, ARPA

[4]P. Baran, "On Distributed Communications: Introduction to Distributed Communications Networks," Rand Report RM-3420-PR, August 1964.

[5]L. Kleinrock, "Information Flow in Large Communication Nets," *RLE Quarterly Progress Report*, July 1961.

was interested in the role of computer technology in the topic of Command and Control Research (CCR).

With the establishment of IPTO, ARPA also keyed on the idea of computers as communications devices rather than just puffed up calculators. In short order, ARPA sponsored a series of network technology projects that led the most important initiative devoted to implementing the concept of a robust, survivable network as envisioned in the earlier Rand study. In 1969, ARPANET was created when four sites in the western United States (the University of California at Los Angeles, SRI of Stanford, California, the University of California at Santa Barbara, and the University of Utah) were connected in a system that was to become the precursor to the Internet.

Bob Taylor, the official in charge of ARPANET, emphasized that ARPANET was mainly about time sharing. Time sharing made it possible for research institutions to share the processing power of large mainframe computers for large calculations. But sharing of computing resources, though important by itself, was not the feature of ARPANET that was to secure its place in history.

From the start, ARPANET was a great success. Not only did it demonstrate the basic objectives of a robust and survivable network, it immediately became a tool for communications, collaboration, and resource sharing by connecting researchers in academic and government laboratories. Over those first few years, ARPANET developed and expanded as new sites were added, and as new capabilities were introduced.

In the 1970s, software and protocols were introduced to facilitate email and file transfers within the network. ARPANET became a high speed digital post office as scientists used it to collaborate on research projects. In a demonstration of the law of unintended consequences, ARPANET was rapidly becoming much more than a defensive measure against the threat of nuclear war. And connections were the key. By providing new approaches to communications, and data and resource sharing among the researchers fortunate enough to be part of the network, ARPANET became a stimulus for the accelerating advance of network technology. With the genie out of the bottle, it was only a matter of time before even greater expansion of networking concepts would be realized.

By 1972, the ARPANET system had been publicly demonstrated and an International Network Working Group (INWG) was established. This group, chaired by Vinton Cerf, who now holds Google's whimsical title of Vice President and Chief Internet Evangelist and who is sometimes referred to as the father of the Internet, was comprised of members interested in the exploration of packet-switching concepts.

In the following year, 1973, ARPANET expanded into a global resource when it was connected to University College in London, England and the Royal Radar Establishment in Norway. Vinton Cerf together with Bob Kahn started a project to develop new protocols for network data transfers known as the Transmission Control Protocol/Internet Protocol (or TCP/IP), and thus they coined the term *Internet* for the first time.

VINTON CERF

Born June 23, 1943, Vinton Cerf was a child of the Cold War era. He would become a key player in the development of Cold War driven technology that would change the world. He grew up in southern California and studied mathematics at Stanford University, where he received his B.S. degree, and then he went to work at IBM. His experience in the workplace stimulated the interest he had had since a child in computer and information technology, and he returned to academia for graduate studies under Leonard Kleinrock at UCLA. He earned M.S. and Ph.D. degrees there and became an accomplished computer scientist in his own rite. In fact, he is now considered to be one of the founding fathers of the Internet. What did he do to earn that accolade? It was his influential role in the creation of the Internet and perhaps most importantly the development of the TCP/IP protocols that enabled the rapid expansion of this "network of networks."

Packet-switching technology, which was first theorized by UCLA's Leonard Kleinrock as a Ph.D. student at MIT in 1961, is the core technology that makes the seamless inter*connection* of geographically dispersed computing systems possible. Kleinrock was convinced that creating a link between two systems without the presence of a physical circuit was possible. He believed that packet switching offered the potential to overcome the limitations of circuit switched networks, as are commonly used for telephone transmission, by enabling "connectionless" linkage through a network.

Although Kleinrock's original concept of networking was sound, significant reliability and addressing limitations arose because the scheme depended on the Network Computing Protocol (NCP). NCP was the initial ARPANET host-to-host protocol, a first version packet-switching concept that had inherently limited addressing capabilities. It's here that Vinton Cerf played a critical role.

Cerf had worked on several networking projects and was a member of the student team at UCLA who staffed the Network Measurement Center, an ARPA-sponsored project that anticipated ARPANET. Toward the end of 1968, Cerf became part of a group of students working on the roll-out of the ARPANET at its first four nodes. The student group, called the Network Working Group (NWG), proved to be valuable in anticipating many of the issues and problems that would arise when ARPANET was initiated.

In 1973, when Cerf had taken a position as professor at Stanford University, he along with Bob Kahn began working to develop the Transmission Control Protocol/Internet Protocol (TCP/IP). TCP/IP was to become the standard basis for data transmission over the Internet. The concept included the idea of *gateways* for directing and routing packets, error correcting processes, and global addressing. In September 1973, Cerf and Kahn convened a working group to discuss the development of the protocol. In 1974, they completed the paper "A Protocol for Packet Network Intercommunication",[6] which set much of the foundation for internet data transmission, and published it in the *IEEE Transactions on Communications*.

Cerf along with graduate students Yogen Dalal and Carl Sunshine published the technical specification of TCP/IP in December 1974. TCP/IP became the foundational protocol for the concept of the Internet; these protocols enabled the connection of separate networks into a network of networks; and the Internet is widely recognized to be fundamentally a global network of networks.

Subsequently, in 1976, Cerf moved to the Defense Advanced Research Projects Agency (DARPA), the successor agency to ARPA. In his new position, Cerf continued to play the key role in the development and establishment of TCP/IP.

While many of the advances of the 1970s were below the radar screen in terms of public awareness and publicity, that was all to change in the 1980s. In 1983, TCP/IP replaced NCP as the transmission technology for the ARPANET. This removed limitations and opened the way for the final maturing of ARPANET into its successor, the Internet. Following the introduction of the personal computer into the market, the mid-1980s marked a boom in the personal computer and minicomputer industries, while corporations began to use workstations, local area networks, and newly developed network technology to communicate with each other and with their customers.

In 1990, the ARPANET was formally decommissioned, leaving behind the immense network of networks known as the Internet with over 300,000 hosts. Within two years, the Internet had continued the impressive growth it had experienced as the ARPANET, growing threefold to more than one million hosts.

The initial explosive growth of ARPANET/Internet continues to the present day. However, because of this growth, the ability to search items and efficiently use the available information across the Internet has begun to become limited.

TRANSITION TO THE WORLD WIDE WEB

By 1991, three major events forced convergence of technologies and accelerated the development of information technology. These three events were the introduction of the World Wide Web, the widespread availability of the graphical browser, and the unleashing of commercialization.

[6]V. G. Cerf, and R. E. Kahn, "A Protocol for Packet Network Intercommunication," *IEEE Transactions on Communications*, 22(5): May 1974.

While working at the European Particle Physics Laboratory of the European Organization for Nuclear Research, Conseil Européen pour la Recherche Nucléaire (CERN) in Switzerland, Tim Berners-Lee introduced the concept of the World Wide Web in a relatively innocuous newsgroup, alt.hypertext, a distributed discussion newsgroup devoted to information about hypertext and hypermedia.

CERN is the world's largest particle physics research center. It sits along the French–Swiss border near Geneva. Founded in 1954, CERN is well known as one of the premier research establishments in the world.

CERN exists for the purpose of bringing together scientists from around the world to study the fundamental building blocks of matter and the forces that hold them together. As a host organization, it provides its scientists with the infrastructure and tools to conduct their research. These tools include particle accelerators and related laboratory facilities, precision measuring devices, library facilities, computational systems, and office support systems. Collaboration among scientists has always been an important element of any world-class research laboratory, and communication among collaborators can be considered a force multiplyer in the efforts to achieve a "critical mass" of scientific talent.

Tim Berners-Lee had been intrigued with the idea of hypertext, database systems in which objects can be linked to one another. In particular, hypertext refers to texts or documents in which elements can be linked to other elements through hyperlinks. While working at CERN, Berners-Lee and his colleagues laid the foundation for the open standards of the Web with hypertext ideas. They introduced the ideas of the HyperText Transfer Protocol (HTTP) for linking network documents, the HyperText Markup Language (HTML) for formatting Web documents, and the Universal Resource Locator (URL) system for addressing objects on a network.

HTTP is a protocol that enables network browsing by specifying the rules for exchanging files (text, images, sounds, video, or other multimedia files) on a network. HTML is a language for formatting network pages, to include multimedia format. With HTML, the author describes what a page should look like, what types of fonts to use, what color text should be, where paragraphs begin, and many other attributes of the document. A URL (and the related concept of a Universal Record Identifier, or URI) provides the syntax and semantics to specify the location and access of resources on a network. These technologies, taken together, enabled the incorporation of multimedia information on the Internet and provided the practical means for browsing this hyperlinked multimedia information. Taken together, these technologies are considered to be introduction of the World Wide Web. The World Wide Web, in a sense, hijacked the Internet and enabled acceleration in the already prodigious growth rate of that evolutionary network of networks.

Meanwhile, now that the World Wide Web was ready to go, the development of the graphical browser was also coming into being.

In 1992, the audio and video broadcasts over the Internet were introduced as a new and exciting technology. These broadcasts were known as the "MBone," a term derived from the phrase "Multicast Backbone." By this time, the Internet's size had grown to more than 1,000,000 hosts, and the prospects for widespread dissemination of multimedia information were another contributor to the accelerating growth of the Internet.

In 1993, a group led by Marc Andreesen at the National Center for Supercomputing Applications (NCSA), located on the campus of the University of Illinois at Urbana-Champaign, developed the graphical browser, Mosaic. Mosaic is considered to be the first browser generally available to the general public, for use on the World Wide Web. The introduction of browser technology was the second element that acted to unleash the explosive growth of the Web.

As the Internet grew to encompass more and more host sites and the traffic of information access continued to accelerate, studies of traffic began to show signs that not all Web sites on the Internet are "equidistant." Some sites appeared to be acting as hubs and were experiencing a disproportionate share of the traffic as messages were routed through the network. Based on these studies, two important types of Web sites were characterized: hubs and authorities. Authority Web sites are those that provide the most prominent sources of content material, while other sites (i.e., hub sites) serve to assemble guides and resource lists that can direct users to recommended authorities. Clearly, the existence of the hub and authority dichotomy alters the distributed nature of the Web.

In 1994, the World Wide Web Consortium (W3C) was founded under the leadership of Tim Berners-Lee. W3C is comprised of individuals and companies involved with the Internet and the Web. The W3C develops open standards so that the Web can continue to evolve in a coherent way. Its objective is to promote "interoperable technologies (specifications, guidelines, software, and tools) to lead the Web to its full potential."

With the establishment of the technologies that enabled the World Wide Web and the graphical browser, the time was right for a new burst of growth of the Web, and this has certainly taken place as traffic on the Internet expanded at an explosive annual growth rate in excess of 340,000%. When, in 1992, the National Science Foundation (NSF), in a sense the government successor to ARPA through its sponsorship of NSFNet, removed restrictions on commercial usage of the Internet, it's no surprise that the Internet and the World Wide Web became a commercial success.

TIM BERNERS-LEE

Most would agree that no other individual has had a greater impact on the development of the explosive growth of the Internet into the World Wide Web than Tim Berners-Lee. And he continues to be a leader in the evolution of the

Web into an ever more powerful influence on human lives through his work at the World Wide Web Consortium, or W3C.

Tim Berners-Lee was born in London, England in June of 1955. As a baby boomer, it is not surprising that he was destined to make his impact on society by contributing to new technologies that help people connect to one another. His parents were both mathematicians and computer scientists and worked on the Ferranti Mark I, the first commercially produced general purpose computer. He quickly followed in his parents' footsteps by developing a keen interest in computers; at Oxford University, he built his first computer from discarded electronics parts.

Berners-Lee studied physics at Oxford University from which he graduated in 1976; he then began a career in computer science. Between 1976 and 1980 he worked in the telecommunications industry and then, in 1980, he took a position as a software engineer at the European Particle Physics Laboratory CERN, where he encountered the laboratory's complicated information system.

At CERN, Berners-Lee had a chance to consult with his friend, Kevin Rogers, from England. They found that things at CERN seemed chaotic and uncontrolled. The vast experimental hall at CERN was filled with an array of diverse, small experiments. Inside the control room were racks and racks of computing hardware, centrally located control systems for the complex particle physics equipment that was the mainstay of CERN's research.

At the time, in most large organizations, computer resources were centrally located and scientists and engineers revolved around them. Large mainframe computers were run by specialists, and scientists and engineers were detached users of this complex equipment. Most of the scientists and engineers at CERN did not have computer terminals in their offices; they had to come to the terminal room in a central facility to create inputs or receive outputs from computer analyses that they initiated.

But the research environment at CERN was more web-like than hierarchical. The interrelationships among researchers at CERN could be viewed as a web-like set of connections among the ten thousand people on the CERN staff. Only half of these staff members were actually present at CERN at any given time. Many of the others had a desk there but were really visitors from their home institutions. To cope with this chaos of staff interconnections, Berners-Lee took the initiative and wrote a computer program to store the relevant information and enable queries through the use of random associations that he called *Enquire-Within-upon-Everything*, or *Enquire*. This system also provided for links between laboratory documents and publications.

In *Weaving the Web*,[7] Berners-Lee said, "When I first began tinkering with software a program that eventually gave rise to the idea of the World Wide Web, I named it Enquire, short for *Enquire Within upon Everything*, a musty old book of Victorian advice."

[7]T. Berners-Lee, *Weaving the Web*, HarperCollins, New York, 2000.

He continued, "What that first bit of Enquire code led me to was something much larger, a vision, technology, and society. The vision I have for the Web is about anything being potentially connected with anything. It is a vision that provides us with new freedom, and allows us to grow faster than we ever did when we were fettered by the hierarchical classification systems into which we bound ourselves.

"The irony is that in all its various guises—commerce, research, and surfing—the Web is already so much a part of our lives that familiarity has clouded our perception of the Web itself."

He continued, "The Web resulted from many influences on my mind, half-formed thoughts, disparate conversations, and seemingly disconnected experiments. I pieced it together as I pursued my regular work and personal life. I articulated the vision, wrote the first Web programs, and came up with the now pervasive acronyms URL, HTTP, HTML, and of course World Wide Web."

According to Berners-Lee, "there was no 'Eureka!' moment. . . . Inventing the World Wide Web involved my growing realization that there was a power in arranging ideas in an unconstrained, web-like way."

Berners-Lee came to realize that the Web offered the potential not only to store information and create data access and communications channels but also to provide for the linkage, the connection, of information, computers, networks, systems, and people all over the world. In so doing, an individual could have access to all the information stored anywhere on the network, anywhere in the world. "There would be a single, global information space."

Berners-Lee respected the computational role of computers and considered them to be generally useful in performing certain tasks and thereby leveraging the human mind to accomplish bigger and better things. But, in addition, through the World Wide Web, computers could enable the tracking and analysis of connections and the connective relationships that underlie human interactions and the workings of society, thereby revealing entirely new ways of understanding and analyzing the world.

Berners-Lee was aware that other people had pursued similar concepts, but without the follow-through. Two such examples are Vannevar Bush and Ted Nelson.

The idea of hypertext and document linking actually began in 1945 when Vannevar Bush, a computer pioneer, wrote an article[8] for *The Atlantic Monthly* describing a theoretical electromechanical device called Memex. The Memex concept can be described as a system that provided for electronic linkages to enable the display of documents from a microfiche library with automatic cross-references from one work to another. Twenty years later, Ted Nelson (who introduced the term hypertext) drew on Bush's work in proposing a software framework called Xanadu, which had the goal of creating a computer network with a simple user interface.

[8]V. Bush, "As We May Think," *The Atlantic Monthly*, July 1945.

While Berners-Lee's Enquire software proved useful at CERN for cross-referencing information about staff and publications, he did not publish his work. Having left CERN in 1981, he returned in 1984 to work on distributed real-time systems for technical data acquisition and control of systems.

During this second period at CERN, Berners-Lee began to conceive of a different type of Enquire system. He was seeking new ways to simplify the exchange of information. He began to imagine a system that would link all the computers of his colleagues at CERN, as well as those of CERN's associates in other laboratories around the world. To create such linkages via the Internet at this time would not be easy, although by now it had proved to be a reliable networking system. Nevertheless, it was, at this stage of its development, very cumbersome to use.

In any event, research at CERN and other major laboratories was becoming so expensive that it was clear that collaboration was the way of the future. Visiting scientists would conduct their experiments while in temporary residence at CERN, then return home to study their data. Although CERN maintained physical facilities and infrastructures, it was in some sense a virtual laboratory serving an extended community of researchers, many of whom were home based elsewhere. Because of the diversity of backgrounds and home institutions, the scientists were accustomed to using a wide variety of computers, software, and procedures. While it was a tremendously creative environment, the chaos and disorganization that resulted from this diversity of tools created real challenges.

In the Enquire software, Berners-Lee would create typed pages of information about persons, devices, or programs. Each page was considered a *node* in the program, in a similar fashion to an index card. To create a new node, it was necessary to start with a link from an existing node. The links to and from each node would be represented in a numbered list at the bottom of each page, in a similar manner to the listing of references in a paper. To find useful information, it was necessary to browse from the first page.

It is interesting to consider all information in the world as connections. Although a dictionary can be considered the repository of word meanings, words are defined only in terms of other words. In such a case, the structure is everything and what we normally think of as content is less important. In a similar way, computers store information as sequences of characters, so meaning for them is certainly in the connections among characters. Here again, it's the connections that matter.

Enquire stored information using connections, but without using structures like matrices or trees. Enquire ran on CERN's software development computer. It had two types of links: an *internal* link from one page (node) to another in a file, and an *external* link that could jump between files. An internal link would appear on both nodes. An external link went in only one direction.

In March 1989, Berners-Lee submitted a proposal to his CERN supervisors to develop a CERN-wide information system based on the ideas in Enquire.

He received no response. Perhaps this is not surprising, since as a physics laboratory, an information technology initiative did not attract CERN's immediate attention. At CERN, there were committees to decide on appropriate physics experiments, but there was no such process to consider an information technology initiative.

In the meantime, Berners-Lee became increasingly interested in the Internet, and in the concept of hypertext.

By early 1990, Berners-Lee still had received no response to his proposal, but he was able to buy a new kind of personal computer called the NeXT. The NeXT machine had a lot of intriguing features and interesting technology; with the new NeXT computer, Berners-Lee could focus some of his attention on his interest in hypertext. In actuality, he decided to forge ahead on his own to produce a system he called the *World Wide Web*.

He began writing code for the Web on his new computer. The NeXT had great flexibility and the software to create a hypertext program. The NeXT computer allowed him to create applications, menus, and windows easily, just dragging and dropping them into place with a mouse. He wrote a Web client program that would allow the creation, browsing, and editing of hypertext pages. It was basically similar to a word processor.

Next, he had to find a way to turn text into hypertext by distinguishing text that was a link from text that wasn't. Once he overcame this hurdle, he was then able to rapidly write the code for the HyperText Transfer Protocol (HTTP), the language computers would use to communicate over the Internet, and the Universal Resource Identifier, the scheme for document addresses.

By December 1990, he had developed and was using HyperText Markup Language (HTML), which he had written to format pages containing hypertext links.

A browser would decode URIs, while enabling the user to read, write, or edit Web pages in HTML. It could browse the Web using HTTP, although it could save documents only into the local computer system, but not over the Internet.

Berners-Lee then wrote the first Web server—the software that holds Web pages on a portion of a computer and allows others to access them. At long last, he could demonstrate what the World Wide Web would look like. But it worked on only one platform, and an uncommon one at that—the NeXT computer. At this point, the HTTP server was also fairly crude.

A big incentive for putting a document on the Web was to provide universal access to it by any other user. But who would bother to install a client if there wasn't exciting information already on the Web? Getting out of this chicken-and-egg situation was the problem. And Berners-Lee wanted to be able to say that if something was on the Web, then anyone could have access to it—not just anyone with a NeXT computer. Thus it was clear that a global system would need interoperable software and protocols of the type he was developing.

Meanwhile, he took an important step that would demonstrate the concept of the Web as a universal space. He programmed the browser so it could follow links not only to files on HTTP servers, but also to Internet news articles and newsgroups. These were not transmitted in the Web's HTTP protocol, but in another Internet protocol called FTP (File Transfer Protocol). With this move, Internet newsgroups and articles were available as hypertext pages. All at once, a huge amount of content that was already on the Internet was available on the Web.

Nevertheless, it would take a little longer for Berners-Lee to pursue his larger vision of creating a global system. By 1992, there were over a million Internet users worldwide and about 86% of them were in the United States. Through the 1990s, the worldwide number of Internet users grew to over 280 million for a compound annual growth rate of over 74%. By the end of 2000, the United States had nearly 135 million Internet users.

There are numerous factors that were key determinants in the growth of the Internet. Among the most important were low cost Internet access devices and the increasing availability of broadband Internet through technologies such as cable modems and Digital Subscriber Lines (DSLs).

For several years, Berners-Lee improved the specifications of URL, HTTP, and HTML as the technology became popular and spread across the Internet. He left CERN in the early 1990s and spent research stints at various laboratories, including Xerox's PARC in California and the Laboratory for Computer Science at MIT. While many early Web developers became Internet entrepreneurs, Berners-Lee eventually chose an academic and administrative life, starting the World Wide Web Consortium (W3C). He currently directs the W3C, which operates as an open forum of companies and organizations with the objective of leading the Web to its full potential.

Since the idea of hypertext is central to the World Wide Web, the Web is centered on multimedia content characterized by textual data augmented by illustrative inclusions of audiovisual materials. Thus the current focus of the Web is based on the concept of content providers preparing information formatted into Web pages while users access the information by reading or downloading it, and responding to simple information requests such as completion of data forms (e.g., to complete a purchase transaction).

The hyperlinks of the Web represent structures of meaning that extend beyond the meaning represented by the Web page content; however, at present, these Web structures of meaning lack longevity. Furthermore, they tend to be based on retrospective use of information. For example, current search engines at best optimize navigation by taking into account the statistical behavior of Web users.

The Web presently is more of a display of provider content than a means of achieving broad synthesis of diverse information sources or of deriving meaningful relationships between Web contents. In general, the providers have little or no influence on the links to the contents provided by them, and the users have little ability to affect the available content or the paths to its access, unless by becoming content providers themselves.

And searching the World Wide Web can be difficult and frustrating. In the present formulation of the Web, content is provided for direct consumption by human users, and little provision is made for enabling software agents to act automatically or for the latent content of the Web, as represented not only by the directly presented information but also by the linkages or connections among information elements. Providing such improvements would present a major enhancement in the intelligent use of Web resources. Tim Berners-Lee is directing research efforts toward the next generation architecture called the Semantic Web.

PATTERNS OF DISCOVERY

Vinton Cerf wanted to provide global access to networks through packet-switching TCP/IP protocols. He worked in coordination with large organizations that directed much of the work on the ARPANET. There was a distinctive top–down approach to developing this technology.

On the other hand, Berners-Lee worked as a self-motivated individual in the bottom–up style. He wanted to develop an easy graphical hyperlinking capability for documents on a global scale across the Internet initially for his parent organization CERN but later for everyone.

Each inventor faced the problems of creating easy-to-use apparatus for universal access. They succeeded in developing easy protocols and tools: TCP/IP, HTTP, HTML, and URI, which together powered the Internet and World Wide Web.

Vinton Cerf and Tim Berners-Lee each followed a Proof of Principle Pattern.

FORECAST FOR CONNECTING NETWORKS

Internet developers will continue to exploit innovation and develop enhancements to the World Wide Web with the great promise coming from initiatives of the W3C. An emphasis on open standards will fuel the continuing efforts to connect networks to the Web.

In particular, the W3C initiatives to develop the Semantic Web will have a major impact on the Web by bringing interoperability and intelligent applications to the world.

Discoveries Requiring Inspiration and Perspiration

We can expect much activity to exploit, innovate, and develop an improved Internet. The National Science Foundation is sponsoring the development of the next generation Internet, called the Global Environment for Networking (GENI). GENI was established in 2005 as a consequence of the initiation of six grants to the National Science Foundation. GENI will focus on security and is expected to handle the increase in Internet traffic; it will be geared for an increase in content-delivery demands with video.[9]

GENI will go beyond Internet Protocol Version 6 (IPv6) and it will include new naming, addressing, and identity architectures; advanced security architecture; faster rates; traffic documentation; and new services. GENI will enable the vision of pervasive computing by including mobile, wireless, and sensor networks.[10]

Discoveries Requiring New Proof of Principle

The next proof of principle development for the Internet is the completion of the Semantic Web Roadmap.

While search engines cover a huge part of the Web, they return many irrelevant answers. However, search engines that incorporate logic have been limited due to the inability to address the large amount of potential data. The growth of the Web produces a combinatorial explosion of possibilities.

In the future, if it becomes feasible to combine a reasoning engine with a search engine, it may be able to produce more useful results. Such an engine will be able to use logic to weed out all but the correct solutions.

Tim Berners-Lee is leading the W3C initiative on the next generation Web—the Semantic Web. Currently, the W3C team works to develop, extend, and standardize the Web, as well as its languages and tools. The objective of the Semantic Web is to provide an efficient way of representing typed and linked data on the Web, in order to allow machine processing on a global scale. While the Semantic Web technologies are still developing, the future of the endeavor appears bright.

The essential property of the Web is its universality through the power of hypertext.[11] And while today's Web is produced primarily for human consumption, the next generation Web will to a great extent facilitate machine as well as human consumption. At the end of the process we will have databases, programs, and sensor output that can automatically function with little or no human intervention.

The effective functioning of the Semantic Web requires that computers have ready access to structured collections of information and to the inference rules that they require to conduct automated reasoning. Artificial intelligence researchers have long studied such systems and produced today's knowledge representation. Its present state is comparable to that of hypertext before the

[9]"Imagining the Internet: A History and Forecast," Elon University/Pew Internet Project. http://www.elon.edu/e-web/predictions/150/2010.xhtml.

[10]Ibid.

[11]T. Berners-Lee, J. Hendler, and O. Lassila, "The Semantic Web," *Scientific American*, May 17, 2001.

introduction of the Web. Knowledge representation is an important element for many applications, but to fully realize its potential, linkage into a comprehensive global system is necessary.

The Semantic Web will provide a framework that will accommodate not only the reasoning rules but also the data that can be accessed from Web-based knowledge sources. Inference engines, software applications that process knowledge in databases and derive new conclusions, develop new knowledge from existing information. The development of inference engines can play a major role in making sense out of existing information.

To add logic to the Web will require using rules to enable inferences, choose courses of action, and answer questions. A variety of mathematical and engineering issues make this task more complicated. The logic that is implemented must be sufficiently powerful to describe complex properties, but not so powerful that agents can be tricked by being asked to consider paradoxes.[12]

Two important technologies for developing the Semantic Web are already in place: eXtensible Markup Language (XML) and the Resource Description Framework (RDF). XML lets everyone create their own tags. These tags can be used by scripts, or programs, in many ways, but the script writer must know the page writer's intent in writing the page. XML permits users to add arbitrary structure to their documents, but it does not place any requirements on the meaning of the structure.

The Semantic Web won't be possible until software agents have the means to figure out some things by themselves. Fortunately, artificial intelligence gives us two tools to help make this possible. First, knowledge representation is a field that defines how we might represent, in computers, some of what is stored between our ears. Second, inference is a way of using formal logic to approximate further knowledge from what is already known.[13]

With its successful development, the Semantic Web will bring both structure and content to the Web and, in so doing, create an environment where software agents can implement complex tasks for users. The first steps in creating the Semantic Web by integrating it into the existing Web are already under way. In the near future, we can expect that these developments will provide new functionality as machines become better able to "comprehend" and process the data.

[12]Ibid.

[13]U. Ogbuji, "The Languages of the Semantic Web," *New Architect*, June 2002.

7

Connecting Devices

Ubiquitous computing represents a powerful shift in computation, where people live, work, and play in a seamlessly interweaving computing environment.
—Mark Weiser[1]

Imagine living your life within the confines of the region that surrounds you. You could call this region your "personal space"—the physical space that surrounds you. As you travel from place to place, this designed region travels with you just like a "bubble" connecting you to other people, places, and things. If you look around this space, what kinds of electronic devices would you find and how would they function?

With the introduction of wireless communications and networking technologies, we are beginning to see the impact of readily *connecting devices* to all aspects of our activities. Our cell phones can now connect us to the Web, as well as to other people. And this technology also demonstrates the new mobility of wireless appliances. As devices with microprocessors permeate our world and become interconnected, we see the dawn of the era of ubiquitous computing. Beyond the interconnection of devices (ubiquitous computing), we can also anticipate two subsequent trends simultaneously emerging: the introduction of Web control of ubiquitous devices, or the ubiquitous Web; and the introduction of artificial intelligence on the Web, offering ubiquitous intelligence.

In this chapter, we tell the story of Mark Weiser and Jeff Hawkins, key players in the introduction of networking and small device technology, and how the ability to wirelessly connect our devices is leading the way toward ubiquitous computing. The associated stories of the ubiquitous Web and ubiquitous intelligence follow naturally in the subsequent chapters.

[1]M. Weiser, "The Computer of the 21st Century," *Scientific American*, September 1991.

Connections: Patterns of Discovery By H. Peter Alesso and Craig F. Smith
Copyright © 2008 John Wiley & Sons, Inc.

THE UBIQUITOUS COMPUTING STORY

Digital devices touch our lives every day. Not only are we inundated by a diverse selection of new electronic products—for example, digital cameras, video games, mobile telephones, MP3 players, PDAs, to name a few—but increasingly we find that ordinary devices not normally considered "digital"—automobiles, toasters, door locks, dishwashers, lawn watering equipment, and the like—are incorporating digital features. In effect, we are surrounded by a dazzling array of electronic devices, and the number and diversity of these gadgets are increasing rapidly.

Perhaps we can get a glimpse of this future when we consider the phenomenon of the cell phone. Even today, many people maintain near continuous contact with their friends and business associates through this mobile communication device. And with mobile Web capabilities being introduced for new cell phones, people are increasingly able to communicate, access information, and transact business on the fly from their mobile personal space. Although we may occasionally experience dropped calls, dead space, and poor quality reception, we can still begin to see the promise of maintaining communication and connection as we move about from place to place; in effect, we have a limited version of our designed personal space.

Add to this the technology of wireless networking that has become commonplace in our homes and businesses of late. First, this technology has allowed us to avoid the tangle of wires that has traditionally been a part of our local computer systems. However, perhaps more importantly, the rapid introduction of *wireless hotspots* in public libraries, businesses, consumer shops (such as Starbucks), airplanes, and, in some cases, whole cities has begun to enable a new type of wireless mobility offering high quality full-featured connectivity at an increasing number of places we may visit.

These two trends—the rapid proliferation of electronic devices permeating all the regions of the globe and all the aspects of human environment; and the accelerating trend toward device interconnection and especially wireless, mobile applications—combine to create the potential for a new paradigm in computing—that of ubiquitous computing.

And where does this concept lead us? In addressing the idea of global interconnection of communications and computing resources, we can consider three phases: (1) connecting devices, leading to ubiquitous computing; (2) the introduction of Web control of ubiquitous devices, leading to the ubiquitous Web; and (3) the ultimate expansion of artificial intelligence on the Web, leading to ubiquitous intelligence. While ubiquitous computing is addressed in this chapter, ubiquitous Web and ubiquitous intelligence are considered in the subsequent chapters.

The idea of ubiquitous computing is a natural progression from the early history of computing, where large, centrally operated mainframe computers gave way to the personal computer, small and inexpensive enough that computing could become decentralized and dedicated to the needs of a single user.

But the potential of the personal computer, while it offers great power as an office and computational tool, is multiplied dramatically when it is connected via networking to other computers, equipment, and electronic devices. Furthermore, as the scope of the network of interconnections expands, providing ever broader reach and access to information and also providing paths for ever more wide-reaching communication, we begin to realize the power of the World Wide Web.

We start down the path toward ubiquitous computing by recognizing that the world is being populated with devices using microchips, and we note again that the proliferation of electronic devices is rapid and accelerating. Once we recognize that many of these devices can be interconnected, we can see that this enables the possibility of their control on a wide, potentially global scale. The final step comes when artificial intelligence reaches the capability of managing and regulating devices seamlessly and invisibly within the environment—achieving ubiquitous intelligence. The approach to ubiquitous computing and the consequent assembling of the global network of networks containing much of the world's knowledge constitutes the first step toward Larry Page's quest for perfect search.

It is paradoxical that the more the demands of modern life require us to move about, the more we need to stay in touch with the people and places we leave behind. Wireless devices help keep us connected in this increasingly mobile world.

In the world today, information has become an extremely valuable commodity. However, with billions of devices already in use today, developing multipurpose communications that can receive and transmit compatible signals is a daunting challenge. At the local level, Personal Area Networks (PANs) form device-to-device interfaces at work and at home. At the global level, we must adapt an interlacing complex of networks to connect compatibly to a growing number of possible device-to-device combinations.[2]

Wireless technology can be used to create a network infrastructure called a Wireless Personal Area Network (WPAN). In the hierarchy of networking, which also includes the Wide Area Network (WAN), the Metropolitan Area Network (MAN), and the Local Area Network (LAN), the scope of the PAN and the WPAN is the smallest.

The WPAN in the office workspace offers access to the essential workspace electronic devices, including the desktop computer, the PDA, the printer, mobile phones, hand-held devices, and pagers. In a WPAN, personal devices wirelessly update and connect. And the growth of home automation and smart appliances will require WPAN applications just as in the office.

However, small devices currently suffer from several drawbacks, including slow processor speeds, limited memory, slow wireless connections, and limited battery life. The small mobile wireless computing environment is not able to

[2]H. P. Alesso and C. F. Smith, *The Intelligent Wireless Web*, Addison-Wesley Professional, Boston, 2001.

support large, complex operating systems and applications. Instead, distributed applications, which gain their capabilities from collections of separate devices working in concert, will become the norm. Unlike desktop computers, small mobile wireless devices use a variety of processors and operating systems and are programmed in a variety of languages.[3] This has created a fierce competition to establish the operating standards that would permit flexible interconnectivity.

UBIQUITOUS COMPUTING

Ubiquitous computing has been termed the *third wave* in computing. In this representation, the first wave was the era of the mainframe computer—large, complex computers run by centralized corporate organizations with each computer providing support to large numbers of people. Then, in the second wave, known as the personal computing era, individuals and personal computing machines shared the same desktop. This was the era of decentralized computing. The final phase, or third wave, comes as ubiquitous computing is introduced, creating a work environment in which each person is connected to lots of devices.[4] The key technology to enable this transformation to the third wave is networking, and networking impacts computing at each level from the personal computing level to the global level.

The concept of ubiquitous computing (UC) was first identified by Mark Weiser in 1988 while he was working at the Computer Science Laboratory (CSL) at Xerox's Palo Alto Research Center (PARC). Weiser described UC as a radical approach to achieving the full potential of what computing and networking ought to be.

People's lives are often affected by powerful things that are effectively invisible. This may soon include the impacts from dispersed collections of wireless computing devices that operate in the background and could amount to hundreds of devices per person. Such a decentralized system of processors and related equipment requires the coordination of new operating systems, user interfaces, networks, wireless displays, and other devices. Ubiquitous computing is different from simply providing small computing devices such as PDAs to large numbers of people, because the invisible interconnection of the devices is a key element. The existence of computer and microprocessor systems that are embedded and invisible creates the possibility that we will use them without even being conscious of them, just as we don't think of electricity when we flick on the light switch.

Weiser said, "My colleagues and I at PARC think that the idea of a 'personal' computer itself is misplaced, and that the vision of laptop machines,

[3]Ibid.

[4]"Ubiquitous Computing," Xerox Palo Alto Research Center—Sandbox Server. `http://sandbox.xerox.com/ubicomp/`.

dynabooks and 'knowledge navigators' is only a transitional step toward achieving the real potential of information technology. Such machines cannot truly make computing an integral, invisible part of the way people live their lives. Therefore we are trying to conceive a new way of thinking about computers in the world, one that takes into account the natural human environment and allows the computers themselves to vanish into the background."[5]

MARK WEISER

Mark Weiser was born in Harvey, Illinois in the suburbs of Chicago, in 1952. A product of the postwar era, he was raised during a time when the computing industry was in its infancy and the idea of dedicated personal computers, much less hand-held computers, was far over the horizon. He studied computer and communication science at the University of Michigan and received his M.A. in 1977 and Ph.D. in 1979. He spent the subsequent 12 years teaching computer science at the University of Maryland, College Park.

Following his career as a university professor, Weiser left the University of Maryland and joined Xerox PARC as a research scientist. After his first year at PARC, in 1988 he was named head of PARC's Computer Science Laboratory (CSL). He served as head of CSL until 1994 when he left PARC to found a new company start-up; he returned to PARC two years later to become its Chief Technology Officer.

While at PARC, in 1987, he completed his seminal work in defining the concept of ubiquitous computing. Throughout his career, he had maintained a strong research interest in human–computer interactions. His concept of UC constituted a new approach to the role of computing, an approach intended to take into account the natural human environment and the potential for humans to interact with computers in an almost unconscious fashion. He believed that, eventually, computers and similar devices would disappear into the background and play a meaningful role in human affairs without being an obvious part of day-to-day life. He foresaw the embedding of computers and microprocessors into the human environment in a variety of ways, but in ways that were not at the center of human attention. And he foresaw the benefit of connecting all sorts of devices by networking, which could be successively extended and expanded through the capabilities of the Internet.

Weiser believed that the purpose of a computer is to assist humans in some productive activities, and that the best role for a computer is to perform as a quiet, invisible servant. He thought that acting by intuition was the effective approach, and therefore the computer should extend human capabilities but as a background resource without requiring constant awareness of its presence or actions. He believed in the idea of "calm technology," technology that engages both the *center* and the *periphery* of our attention, while being accepted

[5]M. Weiser, "The Computer for the 21st Century," *Scientific American*, September 1991.

as a conventional part of everyday life.[6] And this could be achieved through UC operating in the background.

During his career at PARC, Weiser's objective was to help fashion the corporation's strategic plan for leveraging the Internet. He believed that we have hardly begun the Internet Revolution, and that the revolution will not be complete until "everything is on the Web. Light switches, pagers, copiers, printers, as well as PCs, benefit from Web connections." [7]

As head of CSL at PARC from 1988 through 1994, Weiser directed the lab in creating several technologies at the core of the next generation Internet, including the development of a new advanced version of Internet Protocol, IPv6; the Resource Reservation Protocol (RSVP), a network-control protocol that enables Internet applications to process multimedia traffic on the Internet; and the MBone, a critical piece of the technology that enables the Internet to become a real-time, multipoint broadcast medium.

It has been observed that UC is, in a sense, the opposite of virtual reality (VR).[8] In VR, humans experience an artificial world through the magic of computer simulation. Humans are put into the artificial world in VR whereas, in UC, computers are put into the human environment. The problems of the two concepts are very different: in VR, the problem is the completion of sufficiently complex computer simulations to create the appropriate sensation of virtual reality. For UC, the problem is the coordination and integration of such diverse considerations as human factors, computer science, engineering, and social sciences.

In considering the embedding of computing resources into a barely visible part of the human environment, an apt analogy is that of the technology of the written word. As a technology, the development of written language was a key milestone in human development. In a way, it can be considered one of the first information technologies. Nowadays, we take this technology for granted, barely realizing that written language permeates our lives and environment: consider, for example, books, newspapers, street signs, billboards as well as written text on our computer monitors. Through the written language, we have ready access to information with hardly more than a glance. Writing is a crucial technology in the human experience, but it has in a sense disappeared into the background.

It is not surprising that a mature technology would be taken for granted as it disappears into the background of our lives; numerous similar examples exist. For example, the electricity we use comes to us through an elaborate infrastructure of power plants, transmission lines, switches, transformers, and soon, of which we are barely aware. Whether we turn on a light switch or start

[6]M. Weiser and J. S. Brown, "Designing Calm Technology," Xerox PARC, December 21, 1995. `http://sandbox.xerox.com/hypertext/weiser/calmtech/calmtech.htm`.

[7]"Xerox Names Computing Pioneer as Chief Technologist for Palo Alto Research Center," Xerox press release, August 14, 1996. `http://www.ubiq.com/weiser/weiserannc.htm`.

[8]A. K. Tripathi, "Reflections on Challenges to the Goal of Invisible Computing," *Ubiquity* 17: May 17–24, 2005.

the garbage disposal, we hardly give a thought to the elaborate production and distribution system that backs up this capability. A similar observation can be made with regard to the complex mechanical machinery under the hood of our cars.

So it should not be a surprise that key technologies, even information technologies, will become taken for granted as they become embedded in our system and society. Even as we become more dependent on them, they become more invisible to us. This is part of the prognosis for UC: ubiquitous computing through embedded resources that become progressively more invisible to us even as they become more important in our lives.

Mark Weiser died on April 27, 1999, well before he had the chance to see his vision of UC carried out. Nevertheless, he is considered to be a world-class innovator in computer science whose unique vision continues to affect the future of information technology. He was a great pioneer of *ubiquitous computing*, having coined the term in 1988 to describe the wide distribution of computers and microprocessors embedded in everyday objects, and invisible to users.

One of the followers to heed Weiser's vision and attempt to put it into action was a young inventor named Jeff Hawkins, who produced the PalmPilot.

JEFF HAWKINS

Born in Long Island, New York in 1957, Jeff Hawkins is the founder of both Palm Computing in 1996 and Handspring in late 1998, two companies at the forefront of the development of small, hand-held computing devices, originally intended as personal organizers or personal digital assistants (PDAs). Hawkins earned his B.S. degree in 1979 in electrical engineering at Cornell University. In 2005, he became a member of the U.S. National Academy of Engineering, an honor bestowed "for the creation of the hand-held computing paradigm and the creation of the first commercially successful example of a hand-held computing device."

The PDA has great functionality and many diverse uses: current PDAs serve as calculators, clocks and calendars, computer games machines, mobile terminals for accessing the Internet, mobile devices for sending and receiving email, MP3 players, Global Positioning System (GPS) navigation systems, and soon. In fact, the PDA has become the focal point of convergent technology as it has grown from a simple organizer that kept contact and calendar information to a general purpose computing and communications device. Currently, cell phone communications technology and PDA technology are rapidly converging, and many new PDA models have both cell-phone communications as well as mobile Internet access capabilities. As a mobile device with diverse microprocessor and communications capabilities, the PDA should be considered a key technology on the path to UC.

Although he is well recognized for his contributions to the field of decentralized, hand-held computing devices, his interests and contributions did not stop there. With a lifelong interest in the function of the human brain, after founding Palm Computing and Handspring, he created both Numenta, a company developing pattern recognition software, and the nonprofit Redwood Neuroscience Institute, a scientific research institute focused on research related to the human brain.

But the Hawkins success story didn't start in 1996 with the original PalmPilot, described by some as the first fully mainstream PDA device. His interest and success in technology goes back to his upbringing in New York, where he worked with his family in the garage to create a wide variety of seagoing inventions in the mid-1960s.

Jeff Hawkins worked with his father and brothers to design and build boats. He was later to report that this process of invention, hard work, and camaraderie was an influential part of his upbringing. Although he had a great time and learned much about the process of creativity, the boat-building ventures were not financially successful.

While his father and one of his brothers continued in their pursuit of floating platforms, building such structures as a floating hotel, a parking garage, and a floating housing community, Hawkins went on to study engineering at Cornell University, from which he graduated in 1979 with a B.S. degree in electronic engineering. With this educational foundation combined with the creativity and skills he had developed in the family boat-building ventures, he was well trained to go on to build small hand-held computing devices. He was to become a major factor in the revival of the hand-held computer industry, an industry that had received large development resources but that had been faltering before he arrived on the scene.

Following his engineering education, he went to work at Intel Corporation, where, among his other duties, he taught microprocessor design and trained field engineers. In 1982, he left Intel Corporation and took a position at GRiD Systems Corporation. His departure was prompted by his perception that advancement promised to be slow at Intel while the atmosphere at GRiD seemed to offer greater chance for growth and impact.

GRiD Systems was a leader in the development of mobile computing. At GRiD Systems, Jeff's first major accomplishment was the creation of GRiDTask, a high-level programming language for developing mobile applications.

Although he left GRiD Systems to pursue Ph.D. studies in neuroscience at the University of California–Berkeley, he returned to GRiD two years later without the degree but with new knowledge and insights that he immediately put into application. His research at Berkeley led to the development of pattern classifier software that he patented; this software was to become the working prototype for handwriting recognition applications that he pursued at GRiD and subsequently Palm. As Vice President of Research at GRiD Systems Corporation, he developed the GRiDPad, the first hand-held pen-based computer; and the GRiD Convertible, a pen-based tablet computer in which the screen swivels, transforming it into a standard laptop device.

Throughout his career, Hawkins maintained a keen interest in the science of human brain function. His interest in PDAs was, in part, driven by his fascination with the organization of information and its storage and recovery, key elements of both a digital information organizer and the human brain. When he returned to GRiD Systems, he indicated that his goal was "to become famous enough and wealthy enough to really promote and sponsor significant research in neurobiology and theoretical neurobiology."[9]

As he expressed in an interview with MIT's *Technology Review*,[10] Hawkins saw intelligence in terms of the ability of a being to predict the conditions of the surrounding environment based on sensory inputs and logic. He saw that the ability to recognize patterns is a key element of intelligence, and he worked to understand these processes and apply the resulting insights to pattern recognition applications in computer technology. He observed that "the more complex patterns you can predict over a longer time, the more you understand your environment and the more intelligent you are."[11]

The GRiDPad had a simple operating system, a relatively efficient processor, removable storage, and a pen interface. But it was too big and too heavy and had an insufficient battery life. After pursuing some initial concepts in handwriting recognition for the tablet PC, Hawkins decided to switch gears and create a new hand-held device for organizing information. His concept, known as *Zoomer* (a play on the word consumer), a cross between a tablet PC and a PDA, would not be supported by the management at GRiD Systems because of its desire to focus on existing products. So, in 1992, Hawkins left GRiD Systems and founded Palm Computing to work on the Zoomer. He saw great potential in the area of smaller computers. The Zoomer hardware would be developed by Casio and the operating system would be provided by GeoWorks. Palm Computing would put the whole thing together and would integrate the application software, Graffiti.

At the start, Palm Computing was primarily a software company. Hawkins said, "We started Palm Computing with the thought that this would be like the PC world. And the conventional wisdom was that you wanted to be in software. That's where all the big money was being made: Microsoft, Lotus, and Ashton Tate. So we structured it so that Palm would be doing the application software, we partnered with people to do the OS—I selected GeoWorks—we partnered with Tandy to bring it to market, and they brought in Casio to manufacture it, and then we added AOL and Intuit."[12]

[9]S. Barnett, "Jeff Hawkins: The Man Who Almost Single-Handedly Revived the Handheld Computer Industry," *Pen Computing Magazine* 33: April 2000.

[10]"Jeff Hawkins Q&A," *Technology Review*, July 1999. http://www.technologyreview.com/magazine/jul99/qa.asp.

[11]S. Barnett, "Jeff Hawkins: The Man Who Almost Single-Handedly Revived the Handheld Computer Industry," *Pen Computing Magazine* 33: April 2000.

[12]Ibid.

Unfortunately, although the Zoomer had many interesting features, it was slow and had poor text recognition. As a result, the Zoomer did not sell well; however, the Graffiti application was a great success and was subsequently included with a variety of other PDAs, including Apple's Newton.

So Palm went back to the drawing board and refocused its efforts into the creation of software for other handheld computers. Palm also developed *one-touch synchronization*, which eventually became HotSync; and Palm refined its personal information management (PIM) application.

So Hawkins contemplated a new pen system that would fit within the confines of his shirt pocket. His product concept proceeded from his top–down vision of dimension and functionality, which he converted into a physical mock-up. He realized that, to achieve his vision and bring the new product to market, Palm would have to take a comprehensive approach, developing not only the software applications but also the hardware and the operating system. Furthermore, the HotSync technology would be a key element, providing ready connection to allow sharing and updating of data with other computers.

Key to this strategy was making Graffiti successful. Others had previously tried to create handwriting recognition solutions by enabling computers to read the way people write; Hawkins reversed the idea and based his concept on making people learn how to write for the computer.

Hawkins characterized the key attribute of Graffiti to be that it doesn't work like handwriting recognition; it works more like a keyboard. Using the keyboard analogy, he emphasized that it would need to be nearly instantaneous in response; it would have the ability to backspace and introduce punctuation; it would have an editing capability that is much like that of the keyboard.

The result was the PalmPilot organizer and its series of successor devices, collectively known as the Pocket PC. The PalmPilot was one of the most stunning successes in the Information Revolution, leading to the proliferation of small devices, and further ubiquitous computing was launched by Palm. It created a billion-dollar industry virtually overnight and spawned a dazzling set of follow-on products, including Palm III, Palm V, Palm VII, Zire, Tungsten PDAs, Treo smartphones, the LifeDrive and many others, running on both Palm and Windows operating systems.

New related devices are also coming on the scene while capabilities are being added in what can truly be seen as convergence in technology development. MP3 music players are perhaps one of the more visible of the new technologies. Apple's iPod dominated Christmas sales in 2005 when an upgraded model of the iconic music player was outfitted with a video capability, and with several new related products the iPod continues to be a market leader in portable music players. Meanwhile, MP3 and video player capabilities are becoming standard features of new PDA and Pocket PC products, while the distinction between cell phones and PDAs is becoming increasingly blurred. We now have cell phones that can play music and video, provide real-time email and Web access, and perform all the traditional functions of mobile hand-held PDAs, iPhones, and computers. Every day, manufacturers are responding to consumer demands with new systems that make it easier for different devices to work with one another. Everything is connected. Devices are connected to the Web or to each other via a home network or through the telecommunications infrastructure. In the end, convergence is about connectivity. And technology convergence is clearly supporting the progression toward ubiquitous computing.

In 1998, Hawkins left Palm Computing to form Handspring; in part, this was to return to the small, independent company operating environment that Palm Computing had enjoyed before being acquired by U.S. Robotics Corporation, which itself was subsequently acquired by 3-Com. In an interesting series of ironic twists, Palm Computing was to be spun off as an independent operating company in March 2000, and in August 2003, the hardware part of Palm merged with Handspring to form the company PalmOne. In July 2005, PalmOne revised its name and branding, reverting to Palm, Inc.

Jeff Hawkins now serves as a member of Palm's Executive Team with the title of Palm's Founder. His vision of simple, hand-held, mobile computing, and perhaps more importantly his execution of that vision, represents a key contribution toward the concept of ubiquitous computing.

PATTERNS OF DISCOVERY

Both Weiser and Hawkins were focused on making devices universally available from your mobile personal space. With Weiser's vision of ubiquitous computing as a background capability, Hawkins's work on the development of hand-held mobile devices progressively increased capabilities, forming a key achievement of UC.

Their problems included the issues of device compatibility and miniaturization. As they overcame these problems, the result is great progress toward the future of ubiquitous computing by the deployment of relevant ubiquitous devices.

The pattern of discovery they followed was the 1% Inspiration and 99% Perspiration Pattern.

FORECAST FOR CONNECTING DEVICES

To connect devices for ubiquitous computing, we will need to capitalize on the proliferating devices already making their way to every corner of the globe. Ideally, the future wireless communication process will start with a speech recognition user interface, where we merely talk to a personal mobile device that recognizes our words and commands. The personal mobile device would connect seamlessly to embedded and fixed devices in the immediate environment. The message would be relayed to a server embedded in a network with the necessary processing power and software to analyze the message content. The server would link to additional Web resources and draw necessary supplemental knowledge from around the world. Finally, the synthesized message would be delivered to the appropriate parties in their own language on their own personal mobile device.[13]

Discoveries Requiring Inspiration and Perspiration

Let's look at our "personal space" in the year 2020. Starting from your home and traveling through your community to your workplace, what new devices and capabilities will emerge by simply applying inspiration and perspiration?

Input and output capabilities will need to be adapted to the user's current context. For example, it's not a good idea to use speech controls while the user is participating in a meeting. Similarly, direct physical input to a hand-held device when the user is driving a car is not wise.

In addition, when people get together to collaborate their devices should automatically configure to compatible interaction. For example, each person's laptop or PDA should connect and create an association.

Already by the year 2000, the world was populated with some 15 billion devices with microprocessors of many different types. At that time, nearly 30% of all electronic communications were between mobile devices such as cell phones. Current projections are that that will grow dramatically so that, by 2010, 95% of communications will be between mobile devices.

At home we will also find that many of our traditional devices have become commonplace and more useful because of enhanced technology. The three main device categories—PDAs, Smartphones, and hand-held PCs—are offering increasingly easy voice-activated applications and controls.

The PDA market got its start in 1990 with Apple's Newton. Now, with the Smartphone phenomenon only 5 years old, they appear to be overtaking the PDA as the leading category of electronic devices. But a key question is: Will Smartphones decimate the PDA market or will PDAs find new functionality to keep growing? The latter is most likely, because of capabilities such as turn-by-turn navigation, mobile TV, and communication via Bluetooth, WiFi, and WiMax.

Smartphone sales in the year 2005 surpassed 46 million. Smartphone sales can be expected to continue to grow while the potential of this product will depend on what portion of total cell phone sales will become Smartphones. Cell phone sales reached 795 million units in 2005 and will top $1 billion by 2010.

The hand-held PC is just a few years old and its potential is uncertain. However, hand-held PCs have only recently been introduced yet they are projected to reach total worldwide deployment of 200 million units by 2015.

[13]H. P. Alesso and C. F. Smith, *The Intelligent Wireless Web*, Addison-Wesley Professional, Boston, 2001.

All three of these devices can be expected to feature voice-activated commands and networking capabilities to connect us constantly wherever we are.

Radiofrequency identification detectors (RFID) are another interesting device category, and they are already in heavy and expanding use. By 2010, shops may be able to monitor all the items carrying RFID tags you intend to purchase to be processed as you pass out of the door. You will be instantly billed for the items.

By 2012, the very clothing you will wear to work may be equipped with sensors and wearable computers that can detect body warmth and control the temperature of the room.

At home you can expect to find a host of devices serving a great variety of applications. For example, media centers will include Internet delivered television, video on demand over cable, digital video recorders directly off the TV, video game consoles, and toys with built-in tracking technology. Even mood-sensitive home decor along with smart, responsive home and work environments will be available. In addition, your health will benefit from blood-chemistry monitoring, instant electronic identification of pathogens, and windows with coatings to redirect sunlight.

Wireless networks will expand communications to allow people to talk to each other as well as to control the many devices that run our world. Clearly, device-to-device communication must become more efficient and intelligent if we are to realize our expectations of increased productivity. Your personal space will keep in touch with your contacts through upgraded networking capabilities, including new standards for Internet operations and interoperability; open source, peer-to-peer (P2P) interactions; and advanced sensor networks, most with Internet access from mobile devices; and you will use Voice Over Internet Protocol (VoIP) telephony and video conferencing.

As you visit tourist sites in your community, you will find that innovators are developing interactive guidebooks that tourists can use while traveling.

When you get to the office there will be several new capabilities such as e-ink and e-paper. By 2012 these could allow the instantaneous changing of the printed price tags on every item in a store, easy-to-change signage on trucks, and the constant updating of the stories and photos in a newspaper. Paper money will begin to be replaced by smart media.

8

Connecting the Web

The Semantic Web can therefore be thought of as a "smarter," more useful resource.
—Tim Berners-Lee[1]

Over the past forty years, the Internet has grown from an academic experiment in computer networking into the Information Superhighway. The transition from the Internet to the World Wide Web in the 1990s was triggered by the introduction of technologies to enable universal addressing, hypertext transfer, and display of multimedia content. Today we are *connecting the Web* to more and more facets of human activities.

As the Web has become a powerful distribution medium, it has incorporated email, text, images, video, TV programming, and telephone communications. And as we recognize the rapid dispersion of electronic devices throughout the world, we are starting to think about consumer electronics as part of a larger digital ecosystem.

Just as ubiquitous computing is more than simply providing electronic devices to everyone, the ubiquitous Web is more than simply widespread access to the Web. The difference comes about from the enhanced control of the Web as a tool for automated operations. It can enable and control devices dispersed around the world. The Information Revolution will still be getting into high gear as the Web begins to connect light switches, pagers, copiers, industrial machines, as well as PCs.

In this chapter, we tell the story of the ubiquitous Web, with key players Michael Dertouzos and Project Oxygen bringing innovation in the

[1]"Inventor of the Web Explains Next-Gen 'Semantic Web'," *Marketing VOX Daily,* August 3, 2005.

Connections: Patterns of Discovery By H. Peter Alesso and Craig F. Smith
Copyright © 2008 John Wiley & Sons, Inc.

development of concepts for embedded computer systems and their connections; and Tim Berners-Lee working toward automation of the Web through the introduction of Semantic Web architecture. Following the discussion of ubiquitous devices in the previous chapter, these initiatives, aimed toward controlling devices through the Web, constitute the second of the three steps toward achieving Larry Page's perfect search.

The final step will come when artificial intelligence achieves the ability to manage and regulate devices seamlessly and invisibly within the environment. This is called ubiquitous intelligence and we tell that story in Chapter 9.

THE UBIQUITOUS WEB STORY

Today, using application software and widely available computing, a service worker can complete knowledge transactions one hundred times faster than a clerk using written records and manual techniques. As a result, the Information Revolution is placing a 100-fold increase in transaction speed into the hands of the service worker, greatly reducing the cost of information-based transactions. Such productivity gains, which have been well known and respected for some time, stem from both the improved efficiency enabled by software applications and the potential for rapid completion of communications and data exchange by electronic means. But perhaps the most essential ingredient in the Web's continuing success will be search technology, such as Google, which efficiently connects people to relevant information. As a global network of networks, the Web provides large-scale access to information and paths for communications anywhere on the Web; and the access of the information resources embedded in the Web is one of its most beneficial features. Search is a key since it makes the vast information resources usable and accessible.

As microprocessor-based devices become ubiquitous and applications become smarter, how will they affect the Information Revolution and ultimately our world? Two of the ways they will be affected are in the growing ability not only to access information through connected computing and data storage resources, but also to control electronic devices not normally thought of as computing resources; and in the transition from the Web's current role as a repository of relatively static multimedia information for human consumption into information formatted and stored in ways that will be capable of being exploited by automated systems and software agents.

It is clear that the world is seeing an explosive growth in the introduction of new electronic devices, both as innovative technology products—such as iPhones, laptop computers, PDAs, and MP3 players—and also through the introduction of microprocessors into conventional machines in our world that are rapidly being converted into electronic devices. As our world becomes increasingly populated by devices, machines, and appliances that act as information and service sources, their connection to the Web and the standardized approaches to data exchange and control have yet to be worked out. The future Semantic Web architecture, as described by the World Wide Web Consortium (W3C), will link information sources over great distances to provide a common infrastructure to build up services that promote all types of interpersonal and interdevice interactions.

The Web is presently more an exchange conduit and repository of information than a processor of information. In the future, however, the Web will need to do much more than pass raw data between people via search engines.

The ability to put device-understandable data on the Web is becoming a high priority initiative. Tomorrow's software programs should be able to share and process data even when multiple applications and data sources are developed independently. The Semantic Web provides a path for data on the Web to be defined and linked in ways that enable it to be used by other resources, not just for display purposes, but for automation, integration, and reuse of data.[2]

In addition, the dream of an ubiquitous Web entails universal access to relevant capabilities without regard for location, language, or other factors that currently present obstacles to its use. Ideally, the wireless communication process should allow you to speak to a personal device or an embedded device that recognizes your voice, words, and commands. It would connect seamlessly to the correct transmission device, drawing on whatever resources are required from around the Web. Perhaps, in a given request or message, only database search, sorting, and retrieval are required. Or perhaps a specialized application program will be needed. In either case, the information needs will be evaluated and the content of your message, with the appropriate supporting data to fill in the "blanks," will be provided. Finally, the results will be delivered to the appropriate parties in their own language through their own different and varied connection devices.[3]

To achieve the dream of ubiquitous computing—in which devices of all kinds are connected via the Web—we must overcome problems that are similar to those that Web technologies faced in the past: lack of interoperability and lack of accepted standards. To go further and achieve the dream of the ubiquitous Web—in which automated features are enabled—additional structural problems must be overcome and a new approach to representing information on the Web (i.e., the Semantic Web) must be introduced. Until these problems are overcome, we will not have yet taken full advantage of the huge potential offered by the Information Revolution.

The creation of the smart resources envisioned for the ubiquitous Web depends critically on the introduction of methods for incorporating semantic information as part of the information that is stored on the Web. The current

[2]F. Cervone, "W3C Delivers Standards for the Semantic Web," *Infotoday.com,* February 16, 2004. `http://newsbreaks.infotoday.com/nbreader.asp?ArticleID=16514`.

[3]H. P. Alesso, and C. F. Smith, *The Intelligent Wireless Web,* Addison-Wesley Professional, Boston, 2001.

Web configuration, as a network architecture, is compatible with the vision of a Semantic Web. The pervasive Web infrastructure can readily be expanded to meet the vision of ubiquitous computing since it allows access to all forms of physical devices that are connected to the Web. Through the use of Uniform Resource Identifiers (URIs), achieving this vision will provide information and services that will enrich users' experiences in the physical context, just as the Web does in cyberspace.

There are two aspects of the objective to achieve the ubiquitous Web. The first of these is the visionary approach to ubiquitous computing envisioned by Michael Dertouzos and his brainchild at the Massachusetts Institute of Technology (MIT), Project Oxygen. The second is the work of Tim Berners-Lee, who, after playing the key role in introducing the World Wide Web, has gone on to become the leader of the new effort to develop the proposed next generation Web architecture, the Semantic Web, which extends into the ubiquitous Web.

MICHAEL DERTOUZOS

Michael Dertouzos was born on November 5, 1936 in Athens, Greece. A child during the difficult years of World War II and its aftermath, he had a good upbringing as the son of a Greek Navy admiral and a concert pianist. He graduated from Athens College in 1954 and was then able to attend the University of Arkansas as a Fulbright Scholar, where he earned his B.S. and M.S. degrees. Following his undergraduate and master's degree studies, he attended MIT, where he completed his Ph.D. in 1964, having studied electrical engineering and completed research in the area of threshold logic. He then joined the MIT faculty. In 1974, he was named Director of the MIT Laboratory of Computer Science (LCS).

Under Dertouzos's direction, the LCS made many innovative contributions, including developments related to encryption technology, the spreadsheet, the X Windows system, and the ARPANET. Dertouzos was also instrumental in defining the W3C and bringing it to MIT, where it is hosted at LCS.

Dertouzos was always a true technology visionary. He spent much of his time in deep thought about the future directions of information and innovation. Very early, he foresaw the appearance of personal computing as a pervasive technology. He held patents on a variety of technology innovations, including a graphical display system, an incremental photoelectric encoder, a graphic tablet, and a parallel thermal printer. He wrote numerous books, including *The Unfinished Revolution: Human-Centered Computers and What They Can Do for Us*.[4] He was a dedicated advocate of human-centric computing and a leader of the MIT Project Oxygen. He died unexpectedly in 2001,

[4]M. L. Dertouzos, *The Unfinished Revolution: Human-Centered Computers and What They Can Do for Us,* HarperCollins, New York, 2001.

leaving a great legacy based on his belief that it is essential to bend machines toward human needs, rather than the other way around.

PROJECT OXYGEN

Dertouzos was well aware that computing is becoming more human centered. The trend begun by the introduction of the personal computer as a device for the dedicated use of an individual is continuing as additional features and technologies devoted to the individual user are introduced. In addition, the use of computers and microprocessor devices is spreading everywhere. In the future, under the vision of ubiquitous computing, we will not need to carry our own devices around with us. Instead, reconfigurable generic devices will be available throughout our environment to keep us connected and to bring computation to us, whenever and wherever we might go.

New systems will boost our productivity by performing automated, repetitive human tasks, controlling the diversity of physical devices in our environment, finding the information we need, and enabling us to work together.

Project Oxygen,[5] a project of the MIT Computer Science and Artificial Intelligence Laboratory, is an innovative and forward-looking effort to develop the computing architecture of the future, an architecture that is at the heart of the concept of ubiquitous computing and the ubiquitous Web. The concept of Project Oxygen entails the application of pervasive, human-centered computing through a combination of needed user and system technologies. Oxygen's user technologies are intended to directly address human needs; they include speech and vision technologies that enable users to communicate directly with the Oxygen system in a similar manner to direct interaction with another human being.[6] This idea revolves around establishing a new kind of connection between human users and the ubiquitous computing resources that are embedded in the human environment.

To achieve the vision of Project Oxygen, the project focuses on the development of three key components to the system:

- Hand-held devices
- Embedded devices
- The Network

The hand-held device envisioned by Oxygen is the primary human interface. It would be battery powered and would contain a microphone and

[5]"MIT Project Oxygen: Pervasive, Human-centered Computing." http://www.oxygen.lcs.mit.edu/.

[6]"Privacy and Ubiquitous Network Societies," Background Paper, International Telecommunication Union (ITU) Document: UNS/05, ITU Workshop on Ubiquitous Network Societies, April 2005.

speaker. It would feature a small screen for display of text and pictures, a miniature camera, and an antenna for communications with wireless networks in the surrounding area. This small hand-held device would use mobile software that can change functions whenever there is an available software upgrade to provide a new capability. The device can act as a high speed node on a network when the user enters an office or as a slower node when the user is away from areas of high quality connectivity.

As the user moves about with the hand-held device, and when it cannot find a computer network nearby, the device would shift its function into that of a cellular phone, enabling it to maintain its connection through an alternative wireless network. The hand-held device is diversified in function, as it can be a two-way radio when necessary to talk to other hand-held devices, or it can turn itself into an AM/FM radio or even a television with the right software downloaded.

The hand-held device takes a major step beyond the devices available today—high powered smart cell phones that access the Web. The hand-held device can implement a diversified array of functions with the right software to ensure mobility, functionality, and flexibility.

Today's chips process signals the way city traffic flow systems process cars. Each signal checks at every intersection to see whether it should turn right or left to get to its destination. In the Oxygen hand-held device, software will logically rearrange the internal circuitry so that each signal knows ahead of time all the turns it must make to allow it to zip through its path without having to slow down.

Hand-held devices can accept speech and visual input and reconfigure themselves. For example, when a user uses an anonymous hand-held device, it could customize itself to the user's preferences. The hand-held device couples with a wireless network or other nearby hand-held devices. Hand-held devices utilize the same hardware components as the embedded devices in the Oxygen system, but differ in connections to the physical world, computational power, and software.

Like the hand-held device, the Oxygen embedded devices would be built with special purpose chips and be connected via wireless networks. Embedded devices would accept power-hungry computations from hand-held devices and process them, removing the computational burden from the smaller, mobile hand-held devices. In the home, embedded devices would be connected to heaters, air-conditioning units, telephones, lights, and other appliances and devices. Wall-mounted, touch-sensitive displays with microphones, speakers, and cameras would be connected to embedded devices for system–human interaction.

Oxygen's embedded or stationary devices would be embedded in offices, buildings, homes, transit paths, and vehicles to create continuity in intelligent spaces. Embedded devices include interfaces to camera and microphone arrays, and users would be able to communicate to embedded devices using speech.

The Oxygen Network will be enabled by a set of network protocols that will reside in the hand-held devices and the embedded devices in order to help them cope with mobility, interrogation, collaboration, and adaptation to changes in the environment. The Network protocols are seen as an additional set of capabilities on top of the protocols that handle the Internet.

The Network would connect dynamically and would be capable of changing the configuration of self-identifying mobile and embedded devices to establish the framework for continuous collaboration and communications.

In the Oxygen concept, speech and touch screens would replace the keyboards and mice that are the traditional input devices of current computer systems. The use and recognition of the spoken language is an integral part of the Oxygen concept. Four software components, each with well-defined interfaces, would be designed to interact with each other and with device, network, and knowledge-access technologies.

- The speech recognition component would match acoustic signals against a library of phonemes to convert the user's speech into a sentence of distinct words. This component would rank candidate sentences, either to the language-understanding component or directly to an application.
- The language-understanding component breaks down recognized sequences of words and analyzes them grammatically to properly represent their meaning. It would generate limited vocabularies and grammars from examples.
- The language generation component would build sentences in the user's preferred language.
- A speech synthesizer would convert sentences into speech. The hand-held devices would use downloaded software to reconfigure themselves to perform the necessary communication functions.

With the continuing miniaturization of IC chips, the complex functionality envisioned for the Oxygen components may become realizable. As a consequence of this continuing miniaturization, many new devices can be systems that perform both information processing and wireless communications, clearly a key need for the devices envisioned in Project Oxygen. Theoretically, such devices can be deployed anywhere as smart or intelligent objects.

PERFECT SEARCH

Consider the possibility of querying a search engine with any question and getting a response that is not just a correct answer, but the perfect answer—an answer that takes into account the context and intent of your question; an answer that is aware of who you are and why you are asking. Such a search engine would be capable of incorporating all of the world's accessible knowledge including text, video, and audio. It would be capable of discerning between straightforward requests and more nuanced ones. Following Larry Page, we would call this *perfect search*.

That vision is well removed from the typical search engine of today, but it is the stated goal of most search technology developers including Google, IBM, Microsoft, Yahoo!, and Amazon.

Along the path toward this holy grail of perfect search, the real payoff will occur when all the existing knowledge can easily be searched and organized.

In the near future, as we move from the concept of ubiquitous computing to the concept of a ubiquitous Web, search will move from the current PC-centric Web, to a device-centered Web. With the ubiquitous Web, the telephone, automobile, television, and stereo will be capable of connecting to the web—and all of them will be capable of benefiting from network-aware search.

As electronic devices become increasingly connected to the Web, and as information becomes digitized and processed, we will need navigation aids to cope. For perfect search to happen, search needs to be available everywhere and have access to all relevant information, no matter where it is stored.

This also means that search needs to access the invisible Web, comprised of databases of knowledge that are deeply buried in the Web structure, like the University of California's library system. In general, although there are nearly 100 million books, only several hundred thousand are available in any form online. Some of the missing documents may be digitized but are not yet readily available. To this can be added analog archives of film, television, and published works. It's safe to say that, for the present, some of what could be on the Web is hidden or otherwise inaccessible.

As this additional content becomes accessible, search engines will need to incorporate this new content into their indexes, and that will facilitate moving the world ever closer to the possibility of perfect search.

How will businesses respond to perfect search? In one sense, perfect search is needed to initiate and support the decision process, not to finish the decision-making process. A main trend that will add functionality to search for business purposes is the trend toward increasing the semistructured form of information content on the Web. This is being accomplished generally through the use of meta-tags that describe content. For example, email is semistructured data where "To," "From," and "Subject" fields provide the structure.

Meta-tagging is a means to provide basic structure to data. An originator identifies the elements of a document, such as heading, abstract, authorship, first paragraph, second paragraph, and so on, to moderately improve search results.

How will the Web scale up to provide the infrastructure to meet the ideal process? This is a key question given the rapid expansion of information on the Web. Direct scaling based on manual or brute-force methods is not practical; automated search and reliance on machine intelligence are essential.

One element of Web intelligence is the application of adaptive software and software that is capable of learning. The introduction of such adaptive and learning software will represent a major step forward. This is a critical element for exploring Web intelligence. As a growing portion of the Web incorporates learning algorithms, we could begin to see more intelligent performance.

Learning algorithms are procedures designed to extract knowledge from data through two processes: identifying meaningful patterns, and describing them in a useful and relevant manner. The identification process categorizes or clusters records into subclasses that reflect patterns inherent in the data. The descriptive process summarizes relevant qualities of the identified classes. In machine learning, these two processes are referred to as *unsupervised* and *supervised learning.*

We are currently seeing artificial intelligence (AI) algorithms buried inside client-side tools that perform monitoring, content building, content streaming, and content sharing functions—especially in business and finance. For example, at the server-based end of "intelligent" search services, natural language processing (NLP) and linguistic analysis techniques are used to summarize content and identify relevant entities. *Ask Jeeves* and *Albert* are some of the first crude search engines to use NLP.

TIM BERNERS-LEE AND THE SEMANTIC WEB

We continue with the story of Tim Berners-Lee from Chapter 6 as both the originator of the Web and the leading force behind the effort to formulate the next major initiative, the development of the Semantic Web. Because of these two very important contributions, it is appropriate that his story be addressed twice.

Berners-Lee and the W3C team are currently engaged in a major effort to develop, extend, and standardize the Web's markup languages and thereby design the next generation Web architecture, the Semantic Web.

In his earlier landmark work to develop the World Wide Web, Berners-Lee intended that collaborations among individual researchers would be extended through computers. To extend this capability to include automatic processing of information, machines should become capable of analyzing all the data on the Web utilizing its content, links, and transactions.

Under the vision of the Semantic Web, the day-to-day business transactions could be handled directly through machine-to-machine communications and data transfers. Intelligent "agents" could be deployed to perform many functions, assuming the Web is populated by data in machine-understandable form. To get there from here will require a series of technical advances.

It is important to realize that the dream of a Semantic Web and the conversion to machine-interpretable information handling will require a lot of new work. The Web is far from done.

Today, the Web browser and desktop folders are separate. In the future, we might expect that the role and activities of computers and networks would become invisible. To achieve this, it is necessary that the form of information be independent of where the information is stored. Whether information is in the form of hypertext pages or folders, it should be readily accessible and processable by diverse software applications. File names should become merely another form of URI.

Networks should enable the information space and put their analytical power to work. The first step to creating a Semantic Web in which data can be processed directly by machines is to create the framework for putting data on the Web in a form machines can understand.

To date, there is little of the information on the Web in a form that is machine-understandable, and search engines have not proved to be successful in evaluating documents to convert them into machine-understandable format. It is clear that the documents themselves will need to be fully and easily understood by automated software agents or applications.

The trick to conversion of data is getting a computer to extract information online. One process is known as *screen scraping*—trying to salvage something usable from information that is now in a form suitable only for humans.

The conversion of the Web from a repository of static information to a place where software agents could begin to take on more of the actions now considered the realm of the human being could have profound impact. This is part of Berners-Lee's vision for the Semantic Web.

For the Semantic Web to happen, there will need to be a common language that allows computers to understand data, just as HTML allows computers to display hypertext.

Berners-Lee and the W3C have proposed to develop the Semantic Web architecture by building upon layers of open markup languages. The Semantic Web will support machine-processing capabilities that will automate Web applications and services.

For the Semantic Web to provide intelligent features and capabilities, it will have to trade off the expressive power of new logic languages against the computational complexity of processing large semantic networks. The layered language approach of the W3C seeks to implement a balanced approach toward building the Semantic Web.

Agents on the Semantic Web will perform tasks by seeking information from Web resources while communicating with other Web agents. Agents are software programs that work independently and proactively.

Achieving powerful reasoning with reasonable complexity is the ultimate goal. The challenge is finding the best layering of ontology, logic, and rule markup languages for the Semantic Web that will offer solutions to the most useful Web applications. These include accomplishing important tasks automatically on the Web, such as search, query, and information acquisition for collaborative Web applications and services.

The Resources Description Framework (RDF) model is based on statements made about resources, which can be anything with an associated URI (Universal Resource Identifier). The basic RDF model produces a triple, where a resource (the subject) is linked through an arc labeled with a property (the predicate) to a value (the object).

Ontology is the formal specification of terms within a domain and their relationships. It defines a common vocabulary for the sharing of information that can be used by both humans and computers. Ontology can be in the form of lists of words: taxonomies, database schema, frame languages, and logics. The main difference between these forms is their expressive power. An ontology together with a set of concept instances constitutes a knowledge base.

If a program is designed to compare conceptual information across two knowledge bases on the Web, it must know when any two terms are being used to mean the same thing. In addition, the program should be able to identify common meanings for whatever knowledge bases it encounters. Typically, ontology on the Web will combine taxonomy with a set of inference rules.

RDF is based on the eXtensible Markup Language (XML) and it can be used in files on and off the Web. RDF can also be incorporated in regular HTML Web pages. The RDF specification is a basic one, and it is already adopted as a W3C Recommendation.

To keep a given application simple, RDF documents come with a pointer at the top to its RDF schema master list of the data terms used in the document. Anyone can create a new schema document. Two related schema languages are in preparation, one for XML and one for RDF. Between them, they will tell any person or program about the elements of a Web page they describe—for example, that a person's name is a string of characters, but his age is a number.

The Semantic Web first has to achieve the ability to describe, then to infer, and then to reason. The use of schema is a huge first step, and one that will enable a vast amount of interoperability and extra functionality. However, it still only categorizes data. It says nothing about meaning or understanding.

When people "understand" something new, it means they can relate it to other things they already understand. They see the connections. The Semantic Web envisions a similar approach because it is the basis for how computers can "understand" something. Humans learn at a very early age to associate the word "hot" with a burning feeling. Similarly, when we program a computer to do simple things, like make a bank payment, then we can loosely say it "understands" an electronic check. On the other hand, a computer could complete such a process by following a series of links on the Semantic Web that provide the directions to convert each term in a document into a format amenable to machine processing.

In the current Web, decentralization is the fundamental principle that has resulted in its successful and explosive development; it is likely that the same principle will give the Semantic Web its ability to develop.

The Semantic Web is being designed so that it does not have to answer open questions. That is why it will work and grow. From here on it gets difficult to predict what will happen on the Semantic Web because we will need to be able to define trust boundaries.

The difficulty of semantic search is perhaps its most important limitation. Presently, there are two methods of gaining additional information about documents. The first is to manually create a directory, or portal site, by searching the Web and then categorizing pages and links. The second method is to use automatic Web crawling and indexing systems.

An ultimate goal is to combine a reasoning engine with a search engine, which may actually be able to produce useful results.

The basic intent of the Semantic Web is to bring structure and meaningful content to the Web. It will create an environment where automated software applications can carry out important tasks for users. The first steps in introducing the Semantic Web into the existing Web are already under way. Soon, new functionality will emerge as machines become better able to understand and process the data.

The addition of logic to Web architecture is a complex process but could yield great dividends. Google is already preparing to put millions of library books online. Consider what would happen if that information could eventually be accessed as part of a semantic network with a semantic search engine.

Adding logic and rule systems to the Web will permit a scheme for constructing valid Web inferences. Such an initiative requires that proof systems be formed from sets of rules, which can be chained together to form proofs, or derivations. Through the Semantic Web, logic can be used by software agents to make decisions and search terabytes of data.

PATTERNS OF DISCOVERY

Dertouzos and Project Oxygen were creating a vision of ubiquitous computing through hand-held and embedded devices connected to wireless networks.

Berners-Lee is trying to solve the problem of making the Web machine processable. The issues and problems include creating a reasoning engine based on logic and inference powerful enough to utilize the Internet's store of information such that it would be reliable and trustworthy. The result has been the expansion of the Web toward the Semantic Web—a work in progress.

Dertouzos and Berners-Lee have developed their vision as the pattern of inspiration and perspiration. The proof of principle has yet to occur.

FORECAST FOR CONNECTING THE WEB

While ubiquitous computing is already expanding rapidly, the ubiquitous Web is following at a slower speed. The demanding architecture of the Semantic Web is a top–down design that only yields dramatic results, when it offers access for bottom–up programmers from all over the world to begin to exploit software agents to create control over far away devices and regions. We can expect developers will connect devices to the Web by exploiting the inspiration and perspiration pattern to innovate and develop their ideas.

Discoveries Requiring Inspiration and Perspiration

Inventions based on inspiration can help build the ubiquitous Web. The Web can be utilized in today's environment to transform devices into smart resources, and this process is called *ambient intelligence*. The resulting ubiquitous Web will be a pervasive Web infrastructure, where all devices are resources accessible by URIs. The ubiquitous Web requires that:

- Every resource should have a URI.
- Every resource should be collaborative.
- Agents should be location-aware.
- Resources should be context-aware and user-aware.

The ubiquitous Web would use context-aware tagging to devices. Researchers are currently examining the convergence between Web technologies and ubiquitous computing. An example is the CoolTown project. This project supports the concept of *Web presence* for people, places, and things. In this concept, URIs would be used for addressing and localized Web servers would be used for directories to create a location-aware ubiquitous system. This would create a ubiquitous Web as a net of knowledge for physical objects. The goals of users can be obtained explicitly, and user profiles can be made available throughout the environment to inform devices of user preferences.

9

Connecting Intelligence

Intelligence is the most powerful force in the universe.
—Ray Kurzweil[1]

When the philosopher René Descartes proclaimed his famous observation "Cogito, ergo sum," he demonstrated the power of thought by deriving an important fact (i.e., the reality of his own existence) from the simple act of thinking.[2]

The idea of considering thinking machines, however, is a controversial one; and the concept of applying intelligence to the Web is even more so. Both are ideas that are steeped in the field of artificial intelligence (AI), and both have a long way to go before becoming a possibility. Nevertheless, the Information Revolution is a quest for *connecting intelligence*—wherever possible.

The three steps leading to the creation of ubiquitous intelligence start with ubiquitous computing (discussed in Chapter 7), which populates the world with microchip devices. Then, in the second step (described in Chapter 8), the ubiquitous Web takes control of these devices. Finally, AI is introduced, allowing automated self-management and regulation of devices seamlessly and invisibly within the global computing environment: at that point, we will have achieved ubiquitous intelligence.

In this chapter, we present the story of ubiquitous intelligence by exploring the contributions of three intellectual giants whose work was foundational in bringing intelligence to the Web: mathematician Kurt Gödel, who identified

[1]S. Olson, Interview of Ray Kurzweil, Center for Nanotechnology Responsibility, December 2005. `http://www.crnano.org/interview.kurzweil.htm`.
[2]R. Descartes, *The Philosophical Writings of Descartes (Volume I)*, Cambridge University Press, Cambridge, 1985.

Connections: Patterns of Discovery By H. Peter Alesso and Craig F. Smith
Copyright © 2008 John Wiley & Sons, Inc.

the importance of undecidability in logic systems; logician Alan Turing, sometimes called the father of computer science, who introduced his ideas of logic and computer intelligence through the concepts of the Turing machine and the Turing test; and AI pioneer Marvin Minsky, who presented key concepts of connecting intelligence in all forms. Finally, our discussion of ubiquitous intelligence concludes with exploration of the concept of Web intelligence as the final step in the quest to achieve automated Web interactions and Larry Page's perfect search.

THE UBIQUITOUS INTELLIGENCE STORY

As human beings, we often take the process of thinking for granted, just as we walk, talk, and eat without considering the complexity of those functions. Nevertheless, it's relatively easy to define and understand these other biological functions; thinking is unique. While most would agree that thinking along with the intelligence that it displays constitutes a powerful force in our world and is a distinguishing characteristic of human beings, there the agreement ends.

In much of society's discourse, the term *thinking* is loosely defined and ambiguously applied. This is even more true when we refer to intelligent applications on the World Wide Web.

Ultimately, self-awareness and consciousness are important, if not central, aspects of human intelligence, but these characteristics prove much more difficult to analyze and emulate than other more direct indicators of intelligence, such as the ability to memorize facts or apply methods of logical deduction. In general, *thinking* has more to do with these direct indicators, while *intelligence* refers to the broader array of features of human mental activity. Nevertheless, Descartes connects thinking (or cogitation) with the broad and fundamental feature of intelligence, namely, self-awareness and consciousness.

In the context of modern times, discussion of intelligence frequently relates to the implementation of human-like thinking functions in machines, the subject of AI. In this regard, we can infer some aspects of the concept of thinking by recognizing that we identify an individual as intelligent if he has accurate memory recall, the ability to apply logic, and the capability to expand his knowledge through learning.

In general, thinking can include complex processes that use information, concepts, their interrelationships, and inference or deduction, to produce new knowledge. While human thinking involves complicated interactions within the biological components of the brain, and learning is an important element of human intelligence, AI can be expected to add features that we would call "intelligent" to modern information processing systems, leading to the connection of data to services in more effective ways.

Certainly, the idea of ubiquitous intelligence in the full sense of human-like intelligence will not be achieved any time in the near future. However, if we can develop a concept of machine intelligence, much can be done to begin to approach that ultimate end.

Consider the IBM chess supercomputer Deep Blue. Although Deep Blue successfully defeated the world chess champion Garry Kasparov several years ago, many would say that it did so through brute-force computation and not through the application of intelligence and insight.[3] Nevertheless, the science of AI continues to pursue ambitious attempts to make intelligent machines.

A reasonable expectation might be that computerized devices could achieve a certain level of intelligence in their performance, although without being truly intelligent in the broadest sense. This is why some prefer to use the softer and more flexible term of "smartness" instead of "intelligence." [4] We increasingly see the use of terms like smart applications, smart telephones, or smart software.

A world of ubiquitous intelligence will eventually surround us, as electronic environments become sensitive and responsive to people's needs. Ubiquitous intelligence will include a Web that will provide greatly enhanced convenience, dramatic savings of time and cost, surprising advances in the field of entertainment, and increased safety and security.

Ubiquitous intelligence will improve process automation; bring higher quality products and traceability; protect brands, products, and digital assets; establish new services with user-friendly applications; and provide increased security and stability by means of real-time automatic detection of failures and defects.

Another way of viewing ubiquitous intelligence is in terms of bringing pervasive computational intelligence into the physical world by way of computing, communicating, and smart devices. By smart devices, we mean not just computing equipment (PCs, laptops, PDAs, etc.), but also other devices and appliances that have the ability to access Web services. Such devices may have differing levels of intelligence and may be context-aware, interactive, adaptive, automated, and, in some sense, thinking.

Much will need to be done to bring about ubiquitous intelligence even after we realize the ubiquitous Web, a world of interconnected smart devices capable of being controlled on a global basis through the World Wide Web and of accessing the intelligent support entailed in Web services.

The progression of events that can be envisioned starts with the collection of diverse smart objects that may be interconnected at a local level. These isolated smart spaces would be integrated into increasingly higher levels of smart hyperspaces or hyperenvironments and eventually create a smart global network as they are integrated across the World Wide Web.[5] Taking advantage of continuing trends toward hardware miniaturization, it is clear that the

[3]J. Schaeffer, "A Gamut of Games," *AI Magazine*, September 22, 2001.
[4]J. Ma et al., "A Walkthrough from Smart Spaces to Smart Hyperspaces Towards a Smart World with Ubiquitous Intelligence," 11th International Conference on Parallel and Distributed Systems (ICPADS'05), pp. 370–376, 2005.
[5]Ibid.

new devices will encompass micro- and nanoelectromechanical systems (MEMS/NEMS) technology, thus becoming increasingly inconspicuous in their deployment.

The first phase of the trend toward ubiquitous intelligence started near the end of the 20th century when some test articles of smart objects and initial prototypes of smart spaces were developed. The second phase started at the dawn of the 21st century when research into smart everyday objects, environments, and spaces accelerated, generating some significant improvements for practical applications. One can expect to see more and more work being done on smart devices, smart spaces, and distributed intelligent applications as we move into the next few years. One of the significant implications of the development and proliferation of large numbers of smart devices is that various kinds and levels of intelligence will consequently be deployed, residing in everyday objects, and environments, and dispersed throughout the world.

In our everyday language, we can apply the term *intelligence* to computers, robots, or other machines. However, we frequently mean something quite different from the case of human intelligence. For example, while one might be quite impressed with the intelligence of a child prodigy who can perform difficult arithmetic calculations quickly and accurately, a computer that could perform the same calculations faster and with greater accuracy would not be considered to be particularly intelligent. An individual who has rapid memory recall and who has accumulated sufficient amounts of information to consistently win games such as Scrabble or Trivial Pursuit might also be considered to be very intelligent; while a computer storing much greater quantities of accessible factual information would not. Human experts in chess are normally considered highly intelligent, while computers with chess software, even fairly sophisticated software, are not considered to display the same level of intelligence.

AI is the field of study that considers the nature of intelligence in nonhuman entities and the approaches to developing computer systems capable of intelligent action. Three pioneers who had a profound effect in shaping our concepts of machine intelligence, AI, and the Information Revolution are Kurt Gödel, Alan Turing, and Marvin Minsky.

KURT GÖDEL

In the late 1920s, mathematicians were quite certain that every well-posed mathematical question had to have a definite answer—either true or false. For example, suppose they claimed that every even number was the sum of two prime numbers (Goldbach's Conjecture[6]). Mathematicians would seek to determine the truth or falsity of the claim by establishing a chain of logical

[6]E. W. Weisstein, "Goldbach Conjecture," from *MathWorld*—A Wolfram Web Resource, updated 2007. http://mathworld.wolfram.com/GoldbachConjecture.html.

steps that would lead in a finite number of steps to prove if the claim were either true or false. Most mathematicians believed that such a process would always lead to a definitive result.

But in 1931, logician Kurt Gödel proved that the mathematicians were wrong. Gödel was concerned with the consistency of logic systems. In his work to establish the basis for such consistency, he determined that no logic system can prove itself to be internally consistent. He showed that every sufficiently expressive logical system must contain at least one statement that can be neither proved nor disproved while following the logical rules of that system.

Essentially, Gödel proved that not every mathematical question has to have a yes or no answer. Even a simple question about numbers may be undecidable. In fact, Gödel proved that there exist questions that, while being undecidable by the rules of the logical system, can be seen to be actually true if we jump outside that system. But they cannot be proved to be true within the system.

Born April 28, 1906 in Brünn, Austria-Hungary, Kurt Gödel had rheumatic fever when he was six years old and his health became a chronic concern over his lifetime. He entered the University of Vienna in 1923, where he was influenced by the lectures of Wilhelm Furtwängler. An outstanding mathematician and teacher, Furtwängler was paralyzed from the neck down, and this forced him to lecture from a wheel chair and to use an assistant to write on the blackboard. The relationship with Furtwängler made a great impression on Gödel, who was himself very conscious of his health.[7]

Gödel was also keenly interested in the work of the famous philosopher and mathematician Bertrand Russell, and he studied Russell's book *Introduction to Mathematical Philosophy*, which linked the fields of mathematics and logic. He completed his doctoral dissertation in 1929. His thesis was the proof of the completeness of the first order functional calculus, an important contribution to the areas of mathematical logic and set theory and their role in the foundations of mathematics. He subsequently joined the faculty of the University of Vienna and became a key proponent of the school of logical positivism.

According to O'Connor and Robertson,[8] Gödel is best known for his proof of the *Incompleteness Theorems*. In 1931, he proved that, in any axiomatic mathematical system, there are propositions that cannot be proved or disproved within the limitations of the system. To obtain a proof for such propositions, it becomes necessary to add one or more additional axioms to the system. These insights were fundamental results about axiomatic systems.

Gödel's discovery ended the attempt by mathematicians, which had continued over the previous hundred years, to establish axiom-based logic systems

[7]J. J. O'Connor and E. F. Robertson, "Kurt Gödel," *MacTutor History of Mathematics*, October 2003. `http://www-history.mcs.st-andrews.ac.uk/Biographies/Godel.html`.

[8]Ibid.

that would create a consistent framework for the whole of mathematics. One such major attempt had been the effort by Bertrand Russell and Alfred North Whitehead in their renowned work *Principia Mathematica*. Another was the work by the mathematician David Hilbert to create a foundational formalism for mathematics by showing that all of mathematics would follow logically from a properly chosen finite system of axioms; and that some such axiom system exists and can be proved to be consistent. These efforts were severely discredited by Gödel's results. Gödel's theorem did not destroy the idea of mathematical formalism, but it did demonstrate the limitations on formalism as an approach.[9]

One important consequence of Gödel's results is the implication that a computer, the preferred device for carrying out the lengthy and tedious series of logical steps envisioned by the formal-systems approach, can never be programmed in such a way that it will be guaranteed to deterministically answer every mathematical question that can be posed.

In the early 1940s, after relocating to the United States, Gödel continued his mathematics career, producing additional works of great impact. His paper entitled "Consistency of the Axiom of Choice and of the Generalized Continuum-Hypothesis with the Axioms of Set Theory,"[10] considered to be a classic in the field of mathematics, is a representative example. Gödel held a chair at Princeton University until his death in 1978.

It was the mathematician Alan Turing who translated Gödel's logic results about numbers and mathematics into analogous results about calculations and computing machines.

ALAN TURING

Alan Turing was a great pioneer of the computer science field. His well-known conceptual ideas of the *Turing machine* and the *Turing test* were among the first attempts to characterize the concept of machine intelligence. Turing was a foundational leader in the study of machine intelligence, and his research into the relationships between machines and nature led the way to the establishment of the field of AI. His insights opened the door to the Information Revolution.

Alan Turing was born on June 23, 1912 in London, England. He had a difficult childhood and was separated from his parents for long periods. He struggled through his school years but excelled in mathematics. He studied mathematics as an undergraduate at King's College, Cambridge, from 1931 to 1934.[11]

[9]Ibid.

[10]K. Gödel, "Consistency of the Axiom of Choice and of the Generalized Continuum-Hypothesis with the Axioms of Set Theory," in *Annals of Mathematics Studies*, No. 3, University Press, Princeton, 1940.

[11]J. J. O'Connor and E. F. Robertson, "Alan Mathison Turing," *MacTutor History of Mathematics*, October 2003. `http://www-history.mcs.st-andrews.ac.uk/Biographies/Turing.html`.

In his study of mathematics and logic, he was highly influenced by the works of von Neumann, Einstein, Eddington, and, in particular, Russell's *Introduction to Mathematical Philosophy*.[12]

By 1933, Turing's interest in mathematical logic was beginning to gel. He began to realize that a purely logic-based view of mathematics was inadequate. One of his observations was that propositions in mathematics could possess a variety of interpretations and were not as precisely unambiguous as many mathematicians believed. Turing's work at Cambridge University was centered on probability theory. However, he also focused his attention on the question of mathematical decidability. In 1936 he published a paper, "On Computable Numbers, With an Application to the Entscheidungsproblem." [13] The Entscheidungsproblem was David Hilbert's Decision Problem, the problem in symbolic logic of finding a general algorithm for deciding whether or not certain types of mathematical statements are universally valid. It is clear that Turing was pursuing some of the same lines of research as Gödel.

Turing introduced the idea of a computational machine, now called the Turing machine, which, in many ways, became the basis for modern computing. The Turing machine was an abstract computing machine introduced as a thought experiment to help investigate the limitations of computation. A Turing machine is a *state machine*; that is, it is a machine that can be considered to be in any one of a finite number of states at any given time. The instructions for a Turing machine consist of specified conditions under which the machine will change between one state and another using a precise, finite set of rules (given by a finite table) and depending on the value of a single symbol that it would read from a tape.

A Turing machine includes a one-dimensional, theoretically infinite tape divided into cells. Each cell contains one symbol, which could be either a 0 or a 1. The machine has a read–write head to scan a single cell at a time. This read–write head can move left and right along the tape for successive cell scans.[14]

The action of a Turing machine is determined by (1) the present state of the machine, (2) the symbol in the cell being scanned, and (3) a table of transition rules, which serves as the "program" for the machine. If the machine reaches a condition in which there is not exactly one instruction for a transition of state, then the machine halts.[15]

In essence, the tape constitutes the memory of the machine, and the read–write head represents the mechanism through which data is accessed and results recorded. Two important factors are: (1) the machine's tape is considered to be infinite in length, and (2) one may define a function as

[12]B. Russell, *Introduction to Mathematical Philosophy*, Dover Publications, 1993.

[13]A. M. Turing, "On Computable Numbers, With an Application to the Entscheidungsproblem," *Proceeding of the London Mathematical Society* 2(42): 1936.

[14]D. Barker-Plummer, "Turing Machines," *Stanford Encyclopedia of Philosophy*, first published September 14, 1995; substantive revision November 5, 2004. `http://plato.stanford.edu/entries/turing-machine/`.

[15]Ibid.

Turing-computable if a set of instructions exists that would result in the machine finishing the calculation of the function, regardless of how many steps it takes. A Turing-computable function is one that can be successfully computed using a finite number of steps. These two factors or assumptions ensure that no computable function would fail to be computable on a Turing machine simply because there is insufficient memory or time to complete the computation.

Turing also defined the concept of a computable number to be a real number whose decimal expansion could be produced by a Turing machine.[16] He showed that most real numbers are not computable, even though a countably infinite number of them are. Turing then described a number that is not computable, and he remarked that this seemed to be a paradox since he appeared to have described, in finite terms, a number that cannot be described in finite terms. However, he understood the crux of the apparent paradox: he determined that it would be impossible to decide (using another Turing machine) whether a given Turing machine with a given table of instructions will halt after a finite number of steps or continue on indefinitely. Turing's paper contains ideas related to the "halting problem" that proved to be of fundamental importance to mathematics and to computer science.

In 1939, Turing began working for the British Government in its effort to break the German wartime encryption codes produced by an encryption machine known as Enigma. Together with others, Turing developed the Bombe, a deencryption machine that succeeded in decoding messages sent by the German Luftwaffe. By the middle of 1941, Turing's efforts combined with captured information were instrumental in leading to the decoding of secret German Navy messages, using the first practical programmed computer called Colossus.

In March 1946, Turing proposed a design for the Automatic Computing Engine (ACE). This design provided the basis for modern computers.

Then in 1947, Turing started to explore the concept of intelligent machines. He suggested that a computing machine could be considered to be "intelligent" if it could deceive a human judge into believing that it was human, based solely on its responses to a series of queries. His test—called the Turing test—consists of a person asking a series of questions to both a human subject and a machine. The questioning is done via a keyboard so that the questioner has no direct interaction between subjects, human or machine. A machine with true intelligence will pass the Turing test by providing responses that are sufficiently human-like that the questioner cannot determine which responder is the human and which is not.[17]

Turing was convinced that mathematical problem solving could be reduced to simple steps. These could be used to program computer actions. Turing

[16]P. Taylor, "Alan Mathison Turing (1912–1954)," Australian Mathematics Trust, May 14, 2002. `http://www.amt.canberra.edu.au/turingb.html`.

[17]H. P. Alesso and C. F. Smith, *Developing Semantic Web Services*, A. K. Peters, Ltd., Wellesley, MA, 2004.

considered the logical steps one goes through in constructing a proof to be the same steps that a human mind follows in a computation. He was certain that the ability to solve this type of mathematical problem would be a significant indication of the ability of machines to duplicate human thought.

The Turing machine is the foundation of the modern computer and that returns us to our earlier question about whether a computing machine can have human-like intelligence.

Originally, in suggesting the Turing test, Turing proposed that conversation was the key to judging intelligence. In his test, a judge has conversations (via keyboard) with two subjects, one human and the other a machine. The conversations could be about any subject and would continue for a predetermined period of time (e.g., an hour). If, at the end of this time, the judge could not distinguish between the machine and the human, then Turing argued that we would have to consider that the machine was intelligent.

There are many different views about the utility of the Turing test. Some researchers argue that it is the benchmark test of what is referred to as *strong AI*, the idea that some forms of AI can produce real thoughts and true reasoning. As such, it is crucial to defining intelligence. Other researchers take the position that the Turing test is too weak to be useful in this way, because many different computer concepts could generate sufficiently human-like behavior, but for the wrong reasons.

The controversy surrounding the Turing test is that it doesn't seem to be general enough, and it defines intelligence purely in terms of behavior. Thus the Turing test may not in itself be an adequate test of intelligence. Conversation may not be the most appropriate indicator of intelligence, and real thinking may not be guaranteed by producing human-like sentences, as is easily possible for a computer to be programmed to do.

Let's consider the purpose of building artificial intelligence. Is it to simulate the human mind in order to investigate how it works? Or are we primarily interested in the end result? If we are only interested in the output of a machine's execution of a program, then perhaps the Turing test is directly applicable. In this case, it doesn't really matter how the program created its response, but the fact that the output met the human expectations is enough. The appearance of intelligence could be demonstrated by a program that had merely a large enough database of preprogrammed responses and a good pattern recognizer that could trigger the appropriate output.

A good thought experiment to explore this argument can be found in the Chinese Room Problem of John Searle.[18] In this experiment, it is imagined that a man finds himself in a closed room with a book. A message comprised of Chinese characters is passed to him through a slot in the door. The man refers to the book, which gives him instructions (or rules) to process the messages. He refers to the book of rules and, based on the characters on the

[18]J. R. Searle, "Minds, Brains, and Programs," in *The Behavioral and Brain Sciences, Volume 3*, Cambridge University Press, Cambridge, 1980.

message, it directs him to copy some new Chinese characters onto a piece of paper and pass on the resulting response as his reply to the original message. The man follows the book of rules in preparing the response without understanding a word (or character) of Chinese. John Searle pointed out that a computer program carrying out a similar process wouldn't understand Chinese either; he therefore concluded that computer programs do not have any real understanding of the information they process. He indicated that there was a necessary biological function for true understanding to occur.

In a sense, the Chinese Room can be considered to be analogous to a Turing test since we would not be able to reliably tell the difference between a machine's response and a human's response. In this case, the human is performing a set of lookup actions and process steps that are entirely similar to the steps a computer would take.

Another example of this closed room and communication scheme, introduced by Turing,[19] is known as the Imitation Game. In this game, the subject would sit in the closed room accompanied by a book containing combinations of symbols and their corresponding response symbol strings. When the person outside the room types in a series of expressions, they are transmitted to a screen in the room. When the subject sees the input symbols, she opens the book, looks up the input symbol string, finds the corresponding response string, and transcribes it to her keyboard to provide the response to the person outside the room. After several exchanges of this sort, the person outside the room has no reason to believe that he isn't communicating with someone who thoroughly understands the symbols. But in fact the person in the room has no understanding whatsoever of the meaning of the symbols. She is merely responding mechanically and using the prepared book of correct responses. To her, the strings are as meaningless as they would be to a computer.

In short, what is at issue here is that if the person inside the room has no understanding of what the symbols mean, then by parallel argument, it can be said that the Turing machine doesn't understand the symbols it processes either. And if there is no understanding, then there can be no thinking. Neither the Turing machine nor the person in the room can be said to be thinking because neither actually understands what the string of symbols means. So where is the semantics or understanding? Is it in the machine or the room? In reality, there is no understanding. There is only manipulated symbols.

One point of view is that an observer outside the room would say that the person in the room passed the Turing test by giving correct responses to the input symbols submitted. But another view is that while the subject was sitting inside the room, there was no actual understanding, hence no thought, but only symbol manipulation.

So what does it mean to understand a language such as Chinese? Understanding a language involves being able to translate sentences in that language into an internal conceptual representation and to then reason with the internal

[19]A. M. Turing, "Computing Machinery and Intelligence," *Mind* 49:433–460, 1950.

representation based on a set of preexisting knowledge. Nevertheless, it must be admitted that there is considerable thinking involved in prepackaging the instruction book; in a sense, all the possible (or likely) queries would have to have been anticipated, and appropriate responses prepared. Suppose that we had built an elaborate branching tree or lookup table for a computer instead of an instruction book. Then the computer could have answered all the input queries correctly; the problem with the tree or lookup table structure is that the Turing test is not about the behavior it produces but the way it produces it.

Another way of looking at it is that if you correctly define the "system" as consisting of the combination of the human and the book of instructions, together they form a system that exhibits some form of understanding of Chinese.

In 1950, Turing published the paper "Computing Machinery and Intelligence."[20] This paper is a remarkable work on questions that would become increasingly important as the field of computer science developed. In it, Turing identified many of the problems that today lie at the heart of AI; he laid out the basic ideas of defining machine intelligence by introducing the Turing test, still the acid test for recognizing intelligence in a machine.

The mathematical breakthroughs of Alan Turing and other early computer scientists made the 1950s a time of great optimism about machine intelligence. Researchers believed they could simulate many forms of human reasoning and thought processes. Much of that optimism has proved to be well founded as technologies such as expert systems would embody and manipulate knowledge using symbolic logic and artificial neural networks would be trained to find solutions.

Alan Turing died in 1954.

In many ways picking up where Turing left off, Marvin Minsky has been a leader in the field of AI since the 1950s. Like Turing, Minsky worked on the relationship between computational ideas and human psychological processes and has long been a key player in addressing the question of how to endow machines with intelligence.

MARVIN MINSKY

Marvin Minsky was born in New York City, on August 9, 1927. With the onset of U.S. involvement in World War II, he served in the U.S. Navy from 1944 to 1945. Following the war and his return to civilian life, he began to pursue a career in mathematics, attending Harvard University, where he received a Bachelor of Arts degree in 1950, and Princeton University, from which he received his Ph.D. in 1954.

Minsky has made diverse contributions to technology and, in particular, to the field of AI. In 1951 he built the first neural network machine, known as

[20] Ibid.

SNARC (the Stochastic Neural-Analog Reinforcement Computer). It is considered to be the first neural network learning machine based on random wiring. He also holds several patents for technology inventions such as the first confocal scanning microscope, an optical instrument of exceptional resolution and image quality, and the first head-mounted graphical display.

In 1959, Minsky, along with John McCarthy, founded the renowned AI Laboratory at the Massachusetts Institute of Technology (MIT).

Marvin Minsky's research has led to both theoretical and practical advances in AI and the theory of Turing machines. Minsky is also one of the early pioneers in the development of intelligent mechanical robotics.[21] He designed and built numerous components of robotic systems including the first mechanical hands with tactile sensors; and visual scanners with associated software and computer interfaces. He is a significant leader in the field of robotic technology both within and outside MIT.

In 1961, he published the landmark paper "Steps Toward Artificial Intelligence."[22] This paper brought together the then current state of research in AI, along with an identification of the key problems facing this burgeoning field. In his 1965 paper entitled "Matter, Mind, and Models,"[23] he focused on the topic of self-awareness in machines. In 1969, he and Seymour Papert prepared the book *Perceptrons*,[24] in which they identified the capabilities and limitations of pattern recognition machines.

But by the late 1960s, it was clear that achieving human reasoning in a computer would require Herculean efforts. So AI researchers retrenched and began taking a reductionist approach by breaking down the big problems and addressing the smaller component problems.

In 1974, Minsky published the paper "A Framework for Representing Knowledge,"[25] in which he described a model of knowledge representation for many of the phenomena in the areas of cognition, language understanding, and visual perception. The structure of such knowledge representation, called *frames*, is considered to be an early form of object-oriented programming.

In an interview with *Technology Review*,[26] Minsky said: "What surprises me is how few people have been working on higher-level theories of how thinking works. That's been a big disappointment. I'm just publishing a big new book on what we should be thinking about: How does a three- or four-year-old do

[21]R. M. E. Sabbatini, "The Mind, Artificial Intelligence and Emotions," Universidade Estadual de Campinas, 1998. http://www.cerebromente.org.br/n07/opiniao/minsky/minsky_i.htm.

[22]M. Minsky, "Steps Toward Artificial Intelligence," *Computers & Thought*, 406–450, 1995.

[23]M. L. Minsky, "Matter, Mind and Models," *Proceedings of the International Federation of Information Processing Congress* 1:45–49, 1965.

[24]M. Minsky and S. Papert, *Perceptrons*, The MIT Press, Cambridge, MA, 1969.

[25]M. Minsky, "Framework for Representing Knowledge," MIT-AI Laboratory Memo 306, June 1974. Reprinted in *The Psychology of Computer Vision*, P. Winston (Ed.), McGraw-Hill, New York, 1975.

[26]W. Roush, "Marvin Minsky on Common Sense and Computers That Emote," *Technology Review* July 13, 2006. http://www.techreview.com/Infotech/17164/page1/.

the common-sense reasoning that they're so good at and that no machine seems to be able to do? The main difference being that if you are having trouble understanding something, you usually think, 'What's wrong with me?' or 'What's wasting my time?' or 'Why isn't this way of thinking working? Is there some other way of thinking that might be better?'"

In the early 1970s, Minsky and Papert began developing the theory called *The Society of Mind*.[27] This theory combined insights from the disparate fields of child psychology and AI. The Society of Mind theory proposes that intelligence is an emergent property arising from complex interactions among a variety of agents. The diversity in the sources of intelligent behavior is a consequence of the need to perform different tasks requiring a variety of basically different mechanisms.

UBIQUITOUS INTELLIGENCE

Ultimately the impact of AI contributors such as Gödel, Turing, and Minsky should be viewed in the context of their impact on the path to ubiquitous intelligence. From the perspective of the present time, many factors can affect the deployment of smart objects in real environments; an example is the issue of the cost–performance for attached versus embedded computers. Relatively small and cheap devices such as radiofrequency identification (RFID) tags and sensors usually have limited computational power, memory, and wireless transmission distance. Devices with better performance generally have much higher costs and are much larger in physical size.

Another important factor in considering ubiquitous intelligence is the issue of privacy. Other important factors include device and connection reliability, manageability, and trustability.

In considering ubiquitous intelligence, it is possible to break the components into several categories: smart devices (or objects), smart environments, smart systems, and physical aspects of ubiquitous intelligence. The main considerations and examples for each of these categories are discussed next.

For smart objects, important considerations include the role of embedded software and agents; the use of electronic E-Tags and RFID tags; embedded chips, sensors, and actuators; the impact of miniaturization as reflected in MEMS and NEMS technology developments, wireless transceiver/sensors (called motes), and biometric devices; and smart appliances and wearable devices.

In the category of smart environments, these may consist of rooms, homes, offices, and laboratories; they could be entire buildings, libraries, schools, shops, clinics, and hospitals; geographic areas that may be smart environments could include streets, yards, parks, or entire cities; and for mobile applications, vehicles and highways could be considered smart environments.

[27]"Big Thinkers: Marvin Minsky," published on KurzweilAI.net. http://www.kurzweilai.net/bios/frame.html?main=/bios/bio0023.html?

In terms of smart systems, these might include sensors and intelligent networks; the conceptual frameworks of knowledge representation and ontology; the hardware technologies of wearable computing devices, personal and body area networking systems; software considerations including operating systems, middleware, and intelligent association; and finally intelligent service architecture.

The physical aspects of ubiquitous intelligence include such topics as the system interfaces between real and cyber worlds, end-user interfaces, and their control and programming; and user/object identity and activity recognition.

THE WEB "BRAIN"

The total computing power and data potentially accessible over the World Wide Web is truly enormous. By some estimates, as it continues to grow, it will soon become greater than that of a human brain. In fact, as legacy information is added to the Web while virtually all new information is simultaneously included, it is not too far fetched to consider that much of the totality of human knowledge will eventually reside on the Web.

Note that the architecture of the human brain is closer to the Web than it is to a supercomputer.[28] For one thing, there is no central processor guiding communication and computation on the Web; the processing power is highly decentralized as it is in the human brain. In addition, just as the neurons making up a brain are imprecise, faulty, and die, so too the accessible devices and databases containing contradictory data can come and go on the Web; hardware and software faults and crashes locally do not endanger the power of the Web on a macroscopic level.

Just as brain neurons are richly interconnected and communicate with a simple code of neural electrical potentials, so too the computers on the Web are richly interconnected and communicate using fairly simple protocols and languages such as TCP/IP and HTML.[29] Just as a human brain gets information from the human sensory organs, the Web is becoming connected to an array of sensors of various types.

Most people currently want to be able to search for text, images, or video clips on the Web based on visual content, and to ask questions in natural language. They would benefit from specially prepared summaries of the large and diverse databases on the Web, prepared in such a way as to be much more targeted and relevant than that of a brute force search. These goals will require that improved AI functionality reside on the Web.

[28]D. G. Stork, "Artificial Intelligence in the World Wide Web," published on KurzweilAI.net, March 7, 2001. `http://www.kurzweilai.net/meme/frame.html?main=/articles/art0137.html`.
[29]Ibid.

At the same time, Web content providers, search engine companies, and corporations have a preference for enabling automated transactions. This also implies adding artificial intelligence to their systems.

One of the earliest and most noteworthy distributed intelligence projects is the SETI@home project.[30] In this program, over three million individual computers have contributed the equivalent of 60,000 years of PC computing through a cooperative distributed and coordinated effort. The purpose of this project is the digital filtering of radio-telescope signals in the search for indications of extraterrestrial intelligence. Such an interesting and far-reaching objective has galvanized the support and participation of millions of PC owners in a project that may offer a glimpse of things to come in distributed Web intelligence.

Such raw computational power in distributed form is only the starting requirement for intelligent Web systems. While Moore's Law has shown the necessary growth in hardware, software apparently obeys no equivalent law of improvement. It is hard to argue that software such as the UNIX operating system or even proprietary applications such as spreadsheets, or AI systems such as speech recognizers have improved at anywhere near the rate of hardware over the last two decades.

It is interesting to consider the computing power of the largest supercomputing systems of today. While current supercomputers have the computational power of a fly's nervous system, the software that runs on them has remained a bottleneck, strongly constraining the implementation of most AI systems.

A recognized element needed for the development of AI is data. There is ample proof that lack of data is limiting the development of many AI technologies such as speech recognizers, handwriting recognizers, and reasoners. To a great extent, the right type of data doesn't exist on the Web to fill this void.

If it's not on the Web now, one might ask where a software agent would obtain the information needed for AI applications such as understanding simple sentences. One way is for Web service companies to add that information to the Web; and an example of this might be the company Cycorp, which has embarked on the process of entering such information by hand.

Another way might be to use the Web itself to collect the data contributed by nonexpert Web users or *netizens*; this approach is being taken by the Open Mind Initiative,[31] a global collaborative effort to help develop intelligent software. In this initiative, information is collected from netizens to enable computers to learn and assimilate the general knowledge that we, as human beings, often take for granted.

It appears that the Web itself may provide a much better mechanism for implementing machine intelligence than the linear, centralized supercomputer

[30]Ibid.

[31]"The Open Mind Initiative." `http://www.openmind.org/`.

that AI practitioners have, in the past, expected to be the focus of machine intelligence.

WHAT IS WEB INTELLIGENCE?

AI applications are being vigorously pursued in many fields and are beginning to be used on the Web. AI applications are already becoming useful on the Web, and the future appears highly promising.

In May 1997, one of the world's most interesting chess matches was played. The opponents were, on one side, the reigning World Chess Champion, Garry Kasparov; and on the other side, IBM's Deep Blue Supercomputer. In this tournament, history was made as the match up appeared to be fairly even. For nearly fifty years, AI researchers had dreamed of the chance for such a high visibility demonstration.

Deep Blue was designed to choose its chess move by assessing its possible moves and evaluating the possibilities for countermoves. It was able to identify moves and possible countermoves up to a depth of about 14 levels. By value-ranking the game positions that result from the various move options, using an algorithm prepared in advance by a team of grand masters, the program was able to select its next move in a way that mimicked the intelligence of the grand masters.

One approach to AI is to implement methods of computer science and logic algebras. The algebra would establish the rules. Logic structures have always appealed to AI researchers as a natural entry point. An alternative is to use introspection methods, which observe and mimic the human brain and its behavior, in particular, pattern recognition. Deep Blue was designed to take this latter approach.

The outcome of this six-game match was the decisive victory of Deep Blue over the reigning World Chess Champion by a score of 2-1 with 3 draws. Of course, the victory by itself was not as important as was the attention it brought to the nature of machine intelligence and the realization that chess programming is significant because it uses both logic and introspective methods to simulate human intelligence. The Deep Blue–Kasparov match was much like a real-world Turing test for chess in which the machinc (and AI) won.

Suppose most of a human chess player's skill actually came from an ability to compare the current position against images of 10,000 positions already studied. If the explicitly algorithmic Deep Blue yields essentially the same results as a human, then couldn't the computer and its program be called intelligent too?

However, the current Web consists primarily of static data representations. Search engines are one Web technology designed to automatically process information from large numbers of Web sites to deliver useful processed information. And that's why it's important to consider ways to improve our Web experience and discover the path toward perfect search.

PATTERNS OF DISCOVERY

Kurt Gödel discovered that there are limits to mathematical logic that are relevant to today's computer technology—limitations in logical decidability. This was a discovery that followed the Proof of Principle Pattern. His insights on the limits of axiomatic systems bring us to a better understanding of the limits of scalability of the Web and limits on the quest for perfect search.

Alan Turing also followed the Proof of Principle Pattern in his efforts to lay the groundwork for the underlying fundamentals of digital computing and for the field of AI.

Finally, Marvin Minsky's innovations in AI followed the Proof of Principle Pattern, but crucial further development has been slow to follow.

FORECAST FOR CONNECTING INTELLIGENCE

The steps toward ubiquitous computing and a ubiquitous Web are already being taken and it requires no great stretch of the imagination to visualize the ultimate goal of connecting intelligence. The quest for ubiquitous intelligence, however, is much more demanding and difficult to realize.

Nevertheless, if we break ubiquitous intelligence into three parts, its future appears a little more plausible. First, we consider that advances in computer science and the capabilities of computing systems will continue to develop until something approaching the capability of the human brain might be reached near the year 2025. Second, we consider that AI capabilities will also continue to mature, resulting in the development of isolated knowledge bases that might allow computer hardware/software systems to mount realistic challenges to the Turing test by 2035. Third, we can envision that these AI capabilities will begin to exploit the connections of the Web, resulting in the emergence of ubiquitous Web capabilities by about 2045.

The first step requires continued inspiration and perspiration. The second step requires a new proof of principle innovation, but the final step won't happen without the occurrence of some serendipitous discovery that we aren't aware of at the present time.

We can expect developers will connect intelligent devices to the Web by exploiting the 1% Inspiration and 99% Perspiration Pattern to innovate and develop relevant new innovations in several areas.

Discoveries Requiring Inspiration and Perspiration

The field of computational intelligence can be considered to be a subset of AI and includes soft computing, or software techniques modeled more closely to human reasoning than traditional methods. In coordination with the possible use of software agents, software could be empowered to act on behalf of its human user in a more or less autonomous operation. This field is being studied mainly as an approach to solving complex computational problems.

Among the goals of ubiquitous intelligence is the development of technologies to enable intelligent devices to behave in trustworthy ways, taking into account awareness of themselves and other devices. Such smart devices can exhibit different levels of intelligence. For example, an object with an attached RFID tag may have no real intelligence, but it does have some capability that can contribute to overall systems acting in intelligent ways. Generally, a smart thing represents a different area of potential innovation than Web intelligence, the Semantic Web, or an intelligent cyber world.

Smart things can be considered to come in roughly three categories: smart objects, smart spaces, and smart systems. Smart ubiquitous objects may be very sophisticated pieces of equipment such as smart TVs, cameras, or cell phones. The second category of smart spaces, in reality electronically enhanced real environments, relies on computers that manage and serve smart objects. The third category is that of smart systems, where common services include network communications, traffic management, and environmental systems.

Ubiquitous intelligence embraces the idea of computation everywhere through anonymous and invisible devices. The vision is for computation that is available all the time and everywhere, accessible using speech and leaving it to the computer to locate the necessary and available resources to carry out the required actions. Ubiquitous intelligence would rely on an infrastructure of mobile and stationary devices using self-configuring networks. And stationary devices would be embedded in offices, buildings, homes, and vehicles to create intelligent spaces. They would include interfaces to camera and microphone arrays and users would be able to communicate with the devices using speech. This infrastructure supplies an abundance of computation and communication to be harnessed through levels of software to meet the user's needs.

We can expect that there will be advances to support the transition from our present fixed, wired personal spaces to the flexibility and efficiency of an intelligent wireless Web. Some of the key technology requirements are for wireless devices that are adaptable, the development of protocols for wireless applications, applications to enable wireless small screen displays, and mobile software for devices.

An area of development already under way is that of social linking on the Web. Services such as Google interpret links to a Web page as a peer-endorsement and a machine-readable sign of value. Links have become a currency on the Web.

Another developmental area to expect is holographic television. By 2025 we could be able to watch three-dimensional programming that will provide entertainment, information, and potentially a new communication interface.

In addition to these perspiration/inspiration areas of advancement, we can also expect developers to innovate in the connection of intelligent devices to the Web through the Proof of Principle Pattern in several areas.

Discoveries Requiring New Proof of Principle

AI capabilities will also continue to mature, resulting in the development of isolated knowledge bases that might allow computer hardware/software systems to mount realistic challenges to the Turing test by 2035. In this area, we could expect innovations related to the use of AI technology to mimic the thinking processes of the human brain; the development of capabilities of computer-generated software; and ultimately the development of computer innovations, probably related to advances in parallel processing and neural network systems, leading to human knowledge becoming exceeded by machine knowledge.

Beyond these possibilities are forecasts for discoveries that will require *Serendipity*.

Discoveries Requiring Serendipity

Futurists and technology experts say robots and artificial intelligence of various sorts will become an accepted part of daily life by the year 2050 and will almost completely take over physical work.[32]

To achieve the goal of developing capabilities that will begin to exploit the connections of the Web and the emergence of the ubiquitous Web will require serendipitous discoveries of which we aren't currently aware. It is likely that such developments will be motivated by the approach to a technology singularity such as the acceleration of nanotechnology, robotics, or genetics. The subject of a technology singularity, a concept introduced by inventor and futurist Ray Kurzweil, is the topic of the next chapter.

[32]"Imagining the Internet: A History and Forecast," Elon University/Pew Internet Project. http://www.elon.edu/e-web/predictions/150/2016.xhtml.

10

Connecting Patterns

Ray Kurzweil is the best person I know at predicting the future of Artificial Intelligence.
—Bill Gates[1]

The world is becoming an increasingly interesting place and we humans are the engines of change. Our pursuit of advances in technology in the 20th century has had profound impacts on our society and way of life and, as we progressively approach objectives such as ubiquitous intelligence, we are finding it more and more difficult to adjust to the increasing pace of change. Most especially, technology's rate of innovation is getting faster and faster. If we could measure the accumulated technological progress of the entire 20th century, we would find ourselves on a pace to match or exceed it in just 20 years of the new century, if progress were to continue at today's rate. But even that is an inadequate assumption, since we are experiencing not only increasing change, but also a rate of change that is itself increasing.

In the 1970s, author and futurist Alvin Toffler popularized the concept of "future shock," rooted in the observation that our society is undergoing enormous structural change as it transitions from an industrial to a postindustrial society.[2] His view was that many of the societal problems of the day were related to the issue of coping with an accelerating rate of technological and social change associated with that transition. In the 1980s, Toffler went on to suggest that new societal structures, based on information technology, could be part of the solution to this problem.

[1]A. De Borchgrave, "Commentary: Living Forever," United Press International article, Washington, December 29, 2005. `http://www.upi.com/HealthBusiness/view.php?StoryID=20051229-090610-4704r`.

[2]A. Toffler, *Future Shock*, Random House, New York, 1970.

Connections: Patterns of Discovery By H. Peter Alesso and Craig F. Smith
Copyright © 2008 John Wiley & Sons, Inc.

More recently, renowned inventor, author, and futurist, Ray Kurzweil has offered the concept of a "singularity," change that is so rapid and profound that it threatens the basic "fabric of human history." He has pointed out that, although we traditionally view technological change as a linear process with innovation increasing at a constant rate, the reality is that we are experiencing change not only at a rapid rate but also at a rate that is itself increasing in time. Instead of the linear view, a more appropriate perspective is that technological advances are accelerating with a rate of change that is ever increasing—an exponential rate of growth. And the centerpiece of this accelerating growth is information technology and, ultimately, the prospects for machine intelligence. While Toffler viewed the Information Revolution as the solution to the problem of future shock, Kurzweil paints the more up-to-date picture of it being a force that will bring us to the singularity.

For us, we must consider how *connecting intelligence* through technology can produce acceleration in technology innovation and generate new patterns of discovery. In this chapter, we discuss the story of Ray Kurzweil and his forward-thinking perspectives on accelerating rate of technological change and the singularity. Then we consider the impact of connecting patterns for reviewing the insights from the previous chapters. Finally, we provide some closing comments on the meaning and direction of technological change brought about through connections.

RAY KURZWEIL

Born February 12, 1948, Ray Kurzweil was brought up in Queens, New York. The product of the dawning postwar era of advanced technology and computer science, he was exposed to the pleasures of science fiction from an early age and developed an early interest in computers. It is said that at the age of twelve he wrote his first computer program. But he was not content to use computers for the usual applications of the day.

In 1965, at the age of 17, Kurzweil had his television debut when he appeared as a contestant on the TV program *I've Got a Secret*. In his appearance, he first played a piano composition, and then he revealed his secret to host Steve Allen and to the audience: that he had built the computer that composed the piece that he had just played.[3]

Later, Kurzweil went on to win the first prize in an International Science Fair for this unusual high school accomplishment. This was significant as an early effort to not only build a computer and write its software, but also to understand the mental process of pattern recognition in the field of music. Kurzweil also received recognition for this project as a winner of the Westinghouse Science Talent Search, now called the Intel Science Talent Search, the

[3]E. Frieder and K. Joyce, "Great Ideas: Raymond Kurzweil Receives the World's Largest Award for Innovation," *Spectrum* XIII (3): Fall 2001.

oldest and most prestigious American precollege science contest.[4] Through this experience, he gained insight into the role of pattern recognition in defining human intelligence. With his work combining pattern recognition and computer analysis into a computer-based expert system for music composition, he had also taken the first steps toward developing expertise in one of his lifelong passions: artificial intelligence (AI).

Following high school, Kurzweil went on to university studies at MIT. Once again, he demonstrated his keen insight in computer programming when he used his experience in a college application process to develop software that would assist in matching high school students with appropriate universities. In his second year at MIT, he started a new business based on this software that was, in essence, an expert system for college selection. It proved to be a success, and the company was subsequently sold to a New York publishing company. This experience, like many others to come, enabled Kurzweil to leverage his technical and entrepreneurial talents; in addition, it provided him a great sense of how computers could be used in ways that would prove to be of benefit to ordinary people.

He received his B.S. in computer science and literature from MIT in 1970. In 1974, he formed the first of a series of major business enterprises, Kurzweil Computer Products, Inc. (KCP). This company was established to develop the capability of optical character recognition (OCR) through the use of pattern recognition to identify print characters, regardless of the print quality, style, or font. At the time, the state of the technology in OCR was very limited; it could only be successfully used with special print types.

Kurzweil developed an automatic process for extracting the abstract qualities of each letter shape, and used this pattern recognition capability as the basis for computerized reading of text. Based on this initial developmental work, he went on to introduce the first universal OCR system. In addition, he extended this pattern recognition technology to invent a print-to-speech reading machine for the blind, considered to be the first consumer product to successfully incorporate AI technology, a music synthesizer, and a new concept for a flat bed scanner.

By 1978, the new OCR method pioneered by Kurzweil was being used by LexusNexus, the world's largest database of electronic documents, as part of their news information business. Sometime later, singing superstar Stevie Wonder became aware of Kurzweil's innovations and approached him with the idea of integrating computer-control methods into the studio production of music. This resulted in the 1982 founding of Kurzweil Music Systems with Stevie Wonder as its musical advisor.

Kurzweil went on to found numerous other successful business ventures, publish important AI articles and books, develop new inventions, and become a recognized expert not only in the field of AI but in the more risky area of predicting the future of technology. And here too he has made a major impact.

[4]G. M. Henry, "Can We Talk?" *Time Magazine* 127 (17): April 28, 1986.

With his confidence in science and his certainty in the increasing importance of computer science and AI technologies to enhance human intelligence and ultimately step beyond that goal into the realm of machine intelligence, Kurzweil's ideas have proved to be always interesting and usually provocative. One of his controversial ideas is his willingness to consider the prospect of humans being able to live indefinitely. Whatever the level of controversy in his predictions, so far they have had the unnerving tendency to come true.

According to Kurzweil and his *Law of Accelerating Returns,* we are experiencing an accelerating evolutionary growth in technology.[5] And it is not just computation that is growing exponentially; for example, communications technology parameters such as bandwidth, speed, and price performance are also increasing at rates that amount to doubling every year. As another example, the price performance of DNA scanning for biological research as well as for forensic analysis is also doubling annually. Kurzweil noted: "It took us 15 years to sequence HIV—a huge project—now we can sequence SARS in 31 days and we sequence other viruses in a week." [6]

Kurzweil has assembled substantial data on technology advances and he has showed with his technology-acccleration curves that the longer we use a technology, the more we get out of it. In effect, the advances allow us to use less energy, space, and time to produce the same capacity for less cost. Furthermore, one of his most startling pronouncements is that the future of certain trends is readily predictable. In particular, progress in information technology has turned out to be readily discernible. Even though we may not know specifically how those technology predictions will come about, we can project their magnitude, plan for them, and ignore them at our own peril.

Accelerating rates of advancement are the key to his analysis as summarized in a 2006 conference on computing.[7] He states that "the paradigm shift rate is now doubling every decade, so the twenty-first century will see 20,000 years of progress at today's rate." He went on to say, "The well-known Moore's Law is only one example of many of this inherent acceleration. The size of the key features of technology is also shrinking, at a rate of about 4 per cent linear dimension per decade. Three-dimensional molecular computing will provide the hardware for human-level 'strong' AI well before 2030. The more important software insights will be gained in part from the reverse-engineering of the human brain, a process well under way." He has predicted that, by 2020, a $1000 computer will have the computing power of the human brain.

Why is this happening, and where is it leading us? Kurzweil notes that this change is part of an evolutionary process, where each new step in development

[5]R. Kurzweil, "The Law of Accelerating Returns," published on KurzweilAI.net, March 7, 2001. `http://www.kurzweilai.net/meme/frame.html?main=/articles/art0134.html?`

[6]J. Sutherland, "The Ideas Interview: Ray Kurzweil," *The Guardian*, November 21, 2005.

[7]R. Kurzweil, "*The Coming Merger of Biological and Non Biological Intelligence*," Keynote speech at SC06: The International Conference on High Performance Computing, Networking and Storage, Tampa, FL.

becomes a tool for even greater evolutionary change.[8] And in what may be one of the most provocative and controversial parts of his assessment, he notes the role of evolution to be the cornerstone of the accelerating surge toward ubiquitous intelligence.

In a sense, nature's smartest invention was evolution. Evolution creates a new capability or characteristic and then uses that biological result as the foundation for the next stage of evolution. And for that reason, the next stage is normally more complex. While biological evolution through the well-known process of natural selection is one of the most powerful (or at least important) processes we encounter in the physical world, it is clear that such processes are painstakingly slow; in fact, it took some 4 billion years for human intelligence to emerge on the Earth from this slow and relentless process.

Yet, with the emergence of human intelligence and the establishment of human culture, the floodgates were opened for the much faster process of cultural evolution to take hold. Innovation, technology evolution, and the ingrained drive toward progress experienced by most people have combined to bring about the exponential advances in technological change that we are now experiencing. The brain capacity of the human being must be considered to be a major factor in this process, and cultural evolution has created both the social constructs and the technologies to enhance the value of human intelligence with such advances as writing, the printing press, and modern information technology. Thus we have experienced, in succession but with dramatic acceleration, the impacts of the Agricultural Revolution, the Industrial Revolution, and now the Information Revolution.

EVOLVING COMPLEX INTELLIGENCE

As with other forms of change, human cultural evolution started off at a painfully slow pace. After the emergence of human beings 100,000–250,000 years ago, the use of very simple stone tools was practiced for many millennia with very little change in the human condition. With the introduction of agriculture, probably about 10,000 years ago, a surge in cultural evolution took place as people became more productive, labor became more divided, and information became more valuable to the community as a whole. But human technology continued its relentless and accelerating advance.

A paradigm shift in information access occurred with the invention of the printing press some 550 years ago, and this was followed by the Industrial Revolution that again accelerated the pace of technology change. While the introduction and utilization of stone tools took place over periods of time counted in the hundreds of thousands of years, and the introduction of agriculture took hold over a period measured in thousands of years, we have seen

[8]R. Kurzweil, "*Testimony of Ray Kurzweil on the Societal Implications of Nanotechnology*," Committee on Science, U.S. House of Representatives, Hearing, April 9, 2003.

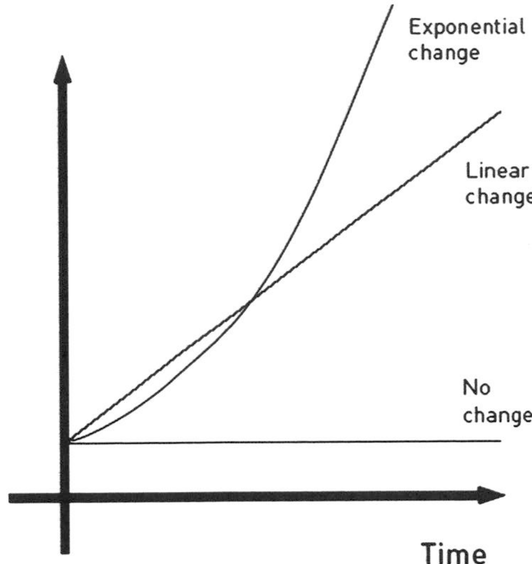

Figure 10-1 *Exponential growth versus linear growth.*

the pace pick up dramatically as we observe the Industrial Revolution and now the Information Revolution of the modern age. And now a paradigm shift, like the development of the World Wide Web, is completed in only a few years' time. The first computers were designed with pencils and drafting paper and built with screwdrivers and wrenches, and today we use computers to create new computers.

Thus the paradigm shift rate is itself growing exponentially, and this is a critical realization. Figure 10-1 compares three types of mathematical relations that can represent different regimes of change. The no change regime displays the stagnant situation. Or it can represent the very early stage of an exponential change, like the use of stone tools by early humans. Linear change is change at a constant rate, the conventional model when thinking about technological change processes. An interesting feature of exponential change is that, while it starts off slowly, it accelerates in time, generating growth at ever faster rates.

Today we find that technology is more and more concerned with knowledge and information, and this has its roots in the transition from the hunter-gatherer society to the agricultural society, where knowledge about the seasons, lunar cycles, rainfall patterns, and soon was much more critical to the success of a stationary agricultural community than it had been previously to migratory groups of hunter-gatherers.

Rolling forward to the modern time, consider the operation of our modern factories. They use software and the Internet to acquire the materials they need at the lowest possible cost. Sophisticated software and computing machinery is also used to arrange for just-in-time delivery of these materials, to plan, manage, and monitor their use in the manufacturing process, to dispatch and route the products when they're finished to minimize delivery costs. And there are only a relatively few people in the modern factory; factories are much less labor-intensive than ever before as automated systems and machinery are leveraged to their fullest. Through the process, inexpensive raw materials are brought to the factory, shaped, converted into high quality products, and dispatched, all with the assistance of sophisticated software. Welcome to the information economy.

Much of the value in our modern economy is generated from the information and knowledge content, where information can be obvious in the product itself (with products such as videos, music, books, software, or databases) or not (as in products such as cars, appliances, processed materials, or food). This effect, like all the others, can be expected to grow exponentially. Increasingly, the knowledge contained in human activities is being stored in electronic form, this too creating an exponential growth in the size of the collective data base. As this knowledge base becomes increasingly available through connections via the World Wide Web, the results will be profound, as will be the problems associated with accessing and processing the data in this burgeoning and massive information source.

With the realization of the reality of accelerating change, we should add to the old axiom "the only things that are certain are death and taxes," that "the only true constant is change."

ACCELERATING RETURNS AND THE SINGULARITY

In the 1950s, the visionary mathematician John von Neumann said that "the ever accelerating progress of technology . . . gives the appearance of approaching some essential singularity in the history of the race beyond which human affairs, as we know them, could not continue."[9] Then a decade later, British statistician and code breaker I. J. Good wrote of the prospects for an explosion in intelligence resulting from the ability of intelligent machines to design new machines without human intervention.[10] This explosion, of course, could be a harbinger of the singularity about which Kurzweil speaks. Good went on to say, "Thus, the first ultra-intelligent machine is the last invention man need ever make." Clearly, if today's computing technologies (e.g., neural nets, genetic algorithms) become mature tools for complex machine design, we can surmise that the singularity is approaching.

The term *singularity* is a reference to the mathematical concept of the same name, which refers to a point in space or time where existing models are no longer valid; an example might be a formula that, at a given point, contains a

[9] J. J. Duderstadt, "*The Future of the University: A Perspective from the Oort Cloud*," Emory University Futures Forum, Atlanta, GA, March 8, 2005.

[10] I. J. Good, "Speculations Concerning the First Ultraintelligent Machine," *Advances in Computers*, Vol. 6, pp. 31–88, Academic Press, San Diego, 1965.

division by zero, an undefined value in mathematical terms. More generally, in mathematics, the theory of singularities is the study of failure in a mathematical structure called a manifold, a term for an abstract mathematical space.

Singularities can occur when the manifold structure degenerates. This represents a breakdown in the fabric of the underlying parameter space itself. Perhaps the most familiar example of this is found in the cosmological theory of general relativity, where the presence of a large mass creates a gravitational singularity that changes the very structure of space–time to the point that it turns back into itself. We call these occurrences black holes: a disturbance in the fabric of space–time, from which nothing can escape, not even light. It is clear that, in this case, it is not just a mathematical construct but a reality that we can observe in the nearby regions of the universe in which we live.

Suffice it to say that, in the study of singularity theory, we frequently encounter situations in which small changes in certain parameters of a nonlinear system can produce large and sudden changes in the behavior of the system. It is in this context that the approach to a technological singularity driven by ever accelerating change warrants thought and study. If we begin to approach a singularity in the technology revolution, it will portend dramatic impacts on the fabric of human society—with results that could be good as well as bad. One area where there can be huge changes in society is in the area of enhanced intelligence.

THE SOFTWARE OF INTELLIGENCE

One of the principal consequences underlying the idea of a technological singularity is the development of the ability to fully understand and reproduce the functioning of the human brain, a topic not only of neuroscience but also AI. If such an objective were successfully achieved, this would have major implications for the field of AI because machines can easily share their knowledge through the process of connections. In theory, if one machine in a network of interconnected machines learns a new skill or fact, this knowledge could be immediately shared with the other machines to which it is connected.

As an example of this, past research on speech recognition has seen efforts in which years have been spent "teaching" a research computer what it needs to know to perform speech recognition. Not only did this entail software development and enhancement, but also the exposure of learning systems to thousands of hours of recorded speech. With each correction, the system improved its performance. Finally, it became reasonably adept at recognizing human speech. Having completed this process the first time, however, the next computer, perhaps a personal computer, would not have to complete the same laborious and time consuming process. It is possible to directly install the software and data to enable fully trained speech recognition immediately.

So what comes next? Kurzweil's vision anticipates that the next steps might start out with the enhancement of the human condition by use of nonbiological materials. One can note that current human intelligence, including memory capacity and brain processing power, is limited by biological realities (i.e., brain size is clearly limited, and brain processing and information transfer speed is limited by the speed of nerve synapses, the biological electrochemical interactions that form the basis for storing, processing, and communicating information). Therefore the possibility of enhanced intellectual functioning by augmentation with nonbiological materials (such as silicon microchips) could offer the prospect of dramatic expansion of the capabilities of the human intellect. And such an idea should not be brushed off as mere science fiction; current technologies such as artificial hearts, pacemakers, and cochlear implants already show that biological augmentation is a current reality, not just a futurist's dream.

Beyond this first step, we can only guess at where things could go from there. Perhaps more important than the immediate impact of the augmentation of physical and mental capabilities by nonbiological materials and devices would be the effect that introduction of such new technologies would have on the evolution of human culture, society, and technology: we could experience another even greater surge in the already rapid and accelerating rate of change. And from there, one can speculate on such possibilities as beings who are not limited by the normal 80–100 year life span or those who utilize software; ultrahigh levels of intelligence in comparison with today's norms; and potentially an expansion of human culture and influence outward into the universe at speeds that are ultimately limited only by the speed of light.

It is through connections of intelligent beings that such a vision could be realized. And through patterns of discovery, the lessons of the past can be projected into the future.

PATTERNS

What does it take to recognize patterns of discovery? In the Introduction, we suggested that recognizing patterns was like becoming a master chess player.

Achieving the status of a chess master consists of three simple steps. First, we need to learn the rules such as the names of pieces, legal moves, chess board geometry, and orientation. Second, we must understand the basic principles including the relative value of the pieces, the strategic value of the center squares, and the power of a threat. Finally, we need to study the games of the masters, including those games containing defining patterns—like the Sicilian Defense.

Similarly, recognizing patterns of discovery requires analogous steps:

- Learn the rules—this requires that talent, knowledge, and resources be skillfully applied.

- Learn the principles—these include serendipity, proof of principle, and inspired exertion.
- Study the designs of masters—find the patterns of inventors such as Edison.

By taking this approach, we can review past and current technology innovations and the scientific patterns of discovery become visible. But understanding recognized patterns is just the beginning of the process of thinking in terms of using patterns and creating new innovation.

In this book, we explored a series of stories about inventors and inventions to examine their patterns of discovery and found insight that we used to forecast elements of the next generation of technology.

We identified three basic patterns of discovery. The first and rarest pattern was the *Serendipity Pattern*, where pure chance produces an accidental discovery. In the second pattern, the *Proof of Principle Pattern,* a scientist starts with a known scientific phenomenon to invent a proof of principle application. The third and most common pattern, the *1% Inspiration and 99% Perspiration Pattern*, reflected the method where a scientist or engineer starts with a known phenomenon and an existing proof of principle application and goes on to invent a new commercially competitive product through a logical progression of advances highlighted primarily by hard work, but fueled by an inspired idea.

Patterns illuminate how collections of discoveries lead to innovation within an entire industry, and discoveries that influence succeeding generations of technology can follow a *pattern of patterns.*

Moore's Law illustrates just such a collection of patterns. The inventions of the vacuum tube, the transistor, and the microprocessor are three essential tributary innovations that feed Moore's Law from its early expression into the present environment and allow us to anticipate future trends. Each of these inventions was the result of a particular pattern of discovery, but together they show a cascade of patterns. In fact, Moore's Law itself also demonstrates the phenomenon of acceleration in succeeding generations of technology as it constitutes an exponential form of growth.

Another aspect of patterns of discovery can be observed when competing discoveries vie for dominance. An example of this is illustrated by the development of the personal computer. The Alto system developed at Xerox PARC followed a Proof of Principle Pattern by incorporating invention of the graphical user interface (GUI), Ethernet, laser printing, and the mouse. PARC may have been the first and best collection of inventions in this area, but it was a commercial failure that left the door open for innovators such as Apple and IBM. The pattern of both the Apple II and the IBM PC was the 1% Inspiration and 99% Perspiration Pattern. Clearly, commercial success can follow a different course than scientific success. And again, the Law of Accelerating Returns can be seen affecting the development of new, competing, smaller personal computing devices.

As we proceed, we search for the "big picture" of where the Information Revolution's innovations in chips, devices, software, and networks are taking us. The end point of the Information Revolution appears to be ubiquitous intelligence, where everyone is connected to devices with access to AI—offering what Google founder Larry Page calls "perfect search."

Our stories have taken us on a journey connecting innovations and inventions that constitutes a quest toward perfect search and ubiquitous intelligence. We can summarize some of the observations in Table 10-1, which lists the inventors, inventions, and patterns discussed in each of the foregoing chapters.

TABLE 10-1 Inventions, Inventors, and Patterns of Discovery

			Pattern of Discovery			
Chapter	Area of Innovation	Key Discoveries	Serendipity	Proof of Principle	Inspiration/ Perspiration	Illustrative Stories
Connecting Information	Search	Efficient ranking			X	Google
Connecting Circuits	Electronics	Edison Effect	X			Moore's Law
		Vacuum tubes		X		
		The transistor		X		
		The microchip		X		
		The computer		X		
		Moore's Law			X	
Connecting Chips	Interactive computing systems	Ethernet		X		The personal computer
		The mouse		X		
		GUI		X		
		The laser printer		X		
		The personal computer		X	X	
Connecting Processes	Software	Computer Languages		X		Software
Connecting Machines	Network	Ethernet		X		Ethernet
Connecting Networks	Packet switching, hyperlinking	TCP/IP/		X		Internet
		HTTP, HTML, URI		X		
Connecting Devices	Ubiquitous computing	Project Oxygen		X	X	Ubiquitous computing
Connecting the Web	Ubiquitous Web	Semantic Web		X	X	Ubiquitous Web
Connecting Intelligence	Ubiquitous intelligence	Artificial intelligence on the Web		?		Ubiquitous intelligence
Connecting Patterns	Patterns of discovery	Law of Accelerating Returns	?			Patterns

CONNECTING PATTERNS

We tend to view the future with a mixture of hope and trepidation—it is filled with uncertainty and consists of one surprise after another. But at the heart of forecasting is the principle of connections. Connections show patterns, and patterns tend to repeat; and this enables a further round of prediction. Not only is accelerating change an observation we can use as a pattern in prediction, the patterns themselves are subject to the law of accelerating change. The future is changing and the act of forecasting changes the future.

Real life in our complex world shows overlapping connections all around us. The consequences of interactions ripple throughout the world. Every action has unintended consequences.

At any given time an invention can offer a new course. Choices are not just individual but collective. What is new with the advent of intelligence is that there can be individual and collective reflection on the process. We start by understanding the process of biological and cultural evolution, and we are then able to reflect on and may be able to influence our own evolution.

The most successful institutions over the long term will be those that not only anticipate near-term shifts but in addition take a longer view. The current emphasis on organizational speed and agility encourages short-term success and survival, but the long-term view is essential to impact the future in a more meaningful way.

PATTERNS OF DISCOVERY

The rapidly accelerating process of change has allowed digital processing to expand in capability to the point that it can significantly magnify and leverage human capabilities to remember, analyze, and predict, thereby creating new connections in succeeding generations of patterns of discovery.

While the introduction of writing ignited the human process of sharing information, other technology advances such as the printing press, the computer, and the Web have progressively added more fuel to that fire, and the acceleration in the impact of information technology has become a dominant force in human society. In particular, the invention of the transistor, the integrated circuit, and digital technology allowed our civilization to take several dramatic steps. Now the Web will become a tool of information sharing.

With the development of the World Wide Web, the works of every author, the knowledge of every individual, and the collective knowledge of humanity are becoming increasingly available to people throughout the world at the touch of a keyboard. It is still necessary, however, to discriminate between the "works of genius and the works of countless monkeys typing at their keyboards."

There is an important difference between innovation designed from the top–down (e.g., major laboratory or university research projects set up by management and directed by national priorities) versus bottom–up (inventors in their garage communicating on the Web). Emergent phenomena from non-directed development may be more amenable to the quick action of the Law of Accelerated Returns than the deliberate top–down initiatives of slow and bureaucratic organizational structures.

The Web demonstrates both bottom–up and top–down phenomena. While the initial development of the ARPANET project was fueled by top–down objectives and planning, when the Internet was born in 1980 as the ARPANET was decommissioned, the bottom–up approach became viable. The wild popularity of email as a new communications paradigm provided impetus for the increasingly bottom–up evolution of the Internet. Then Tim Berners-Lee's innovations that introduced the technology and protocols for the World Wide Web enabled a truly open ended, mostly bottom–up approach that lit the real fire to ignite the explosive growth of the Web.

FORECAST FOR CONNECTING PATTERNS

Kurzweil's view of compressed discovery means that we can expect ever greater innovation feeding directly into cascading generations of progress. Just as Moore's Law expresses the trends leading to the future expansion of hardware technology, a new law of software capacity is required if we are to realize some of the reasonable technologies that could lead us to ubiquitous intelligence. We should anticipate such software breakthroughs within the next decade.

Another phenomenon we can expect is that the time between inspired new devices and capabilities resulting from new proof of principle discoveries will become shorter. We can see this already by considering the transition time on a 10-year time scale for development of the Xerox Alto and the IBM PC with Ethernet, word processing, GUI, and mouse; in contrast, newly developed small devices such as video cell phones and multifunction PDA devices are going from the research lab to commercial production in much shorter periods of about one year. The transition from a proof of principle discovery to inspired spin-offs is clearly accelerating.

EPILOGUE

We began this book by describing our past as a tapestry. We asked how we could extend the tapestry's quilted pattern in order to forecast coming innovations. What we have discovered by unfolding the remarkable stories of inventors is a perspective of the "big picture" for the next generation of information technology—and a respect for the ingenuity of the human race.

Bibliography

Albright, R. E. "What Can Past Technology Forecasts Tell Us About the Future?" *Technological Forecasting and Social Change*, January 2002.

Alesso, H. P. and Smith, C. F. *Thinking on the Web: Berners-Lee, Gödel, and Turing*, John Wiley & Sons, Hoboken, NJ, 2006.

Alesso, H. P. and Smith, C. F. *Developing Semantic Web Services,* A. K. Peters, Ltd., Wellesley, MA, 2004.

Alesso, H. P. and Smith, C. F. *The Intelligent Wireless Web*, Addison-Wesley, Boston, MA, 2001.

Alesso, H. P. and Smith, C. F. "On the Classifying the Deformation Shape of the Liquid Drop Model," *Il Nuovo Cimento*, 1981.

Alesso, H. P. "Elementary Catastrophe Modeling of an End-Loaded Ring in a Rigid Cavity," *Nuclear Engineering and Design*, 1978.

Antoniou, G. and van Harmelen, F. *A Semantic Web Primer*, The MIT Press, Cambridge, MA, 2004.

Associated Press, "Snap.com Plans to Combat Click Fraud," Associated Press online, July 19, 2005.

Baran, P. "On Distributed Communications: Introduction to Distributed Communications Networks," Rand Report RM-3420-PR, August 1964.

Bardeen, J. et al., U.S. Patent 2,524,035, "Three-Electrode Circuit Element Utilizing Semiconductive Materials," Application 1948. Patent to Bell in 1950.

Barker-Plummer, D. "Turing Machines," *Stanford Encyclopedia of Philosophy*, First published September 14, 1995; substantive revision November 5, 2004. `http://plato.stanford.edu/entries/turing-machine/`.

Connections: Patterns of Discovery By H. Peter Alesso and Craig F. Smith
Copyright © 2008 John Wiley & Sons, Inc.

Barnett, S. "Jeff Hawkins: The Man Who Almost Single-Handedly Revived the Handheld Computer Industry," *Pen Computing Magazine* 33: April 2000.

Battelle, J. "The Search: How Google and Its Rivals Rewrote the Rules of Business and Transformed Our Culture," Portfolio 2005.

Bellis, M. "The Inventions of Thomas Edison, History of Phonograph—Lightbulb—Motion Pictures," 2006. Inventors.about.com/library/inventors/bledison.htm.

Bellis, M. "Inventors of the Modern Computer: The History of the Integrated Circuit (IC)—Jack Kilby and Robert Noyce," 2006. www.About.com.

Bennett, D. "The Age of Ray Kurzweil," *The Boston Globe*, September 25, 2005.

Berners-Lee, T. "Semantic Web Primer," 2002. www.w3.org/2000/10/swap/Primer.html.

Berners-Lee, T. *Weaving the Web,* Texere Publishing, New York, 2001.

Berners-Lee, T. *Weaving the Web*, HarperCollins, New York, 2000.

Berners-Lee, T., Hendler, J., and Ora, L., "The Semantic Web," *Scientific American*, 35–43, May 2001.

Bezroukov, N. "Portraits of Open Source Pioneers," 2006. www.softpanorama.org/People/Torvalds/index.shtml.

"Big Thinkers: Marvin Minsky," published on KurzweilAI.net. http://www.kurzweilai.net/bios/frame.html?main=/bios/bio0023.html?

Biography, "Michael Dertouzos," 2004. http://www.kurzweilai.net/bios/frame.html?main=/bios/bio0018.html?

Boden, M. A. "The Social Impact of Artificial Intelligence," in *The Age of Intelligent Machines*, Raymond Kurzwell (ed.), MIT Press, Cambridge, MA, 1992.

Brandow, T. "The Future of Computing Is Invention: An Interview with Alan Kay," Hewlett Packard Feature Story. http://www.hp.com/hpinfo/newsroom/feature_stories/2002/alankay02.html.

Brin, S. and Page, L. *The Anatomy of a Large-Scale Hypertextual Web Search Engine*, Computer Science Department, Stanford University, Stanford, 1996.

Brin, S. and Page, L. "The Future of the Internet," March 21, 2001. www.commonwealthclub.org/archive/01/01-03google-speech.html.

Bronowski, J. *The Creative Process*, Scientific American, W. H. Freeman and Company, New York, 1982.

Burke, J. *Circles: Fifty Roundtrips Through History Technology Science Culture*, Reprint edition, Simon & Schuster, New York, 2003.

Burke, J. *Twin Tracks: The Unexpected Origins of the Modern World,* Simon & Schuster, New York, 2003.

Burke, J. *The Knowledge Web: From Electronic Agents to Stonehenge and Back—And Other Journeys Through Knowledge*, Simon & Schuster, New York, 2000.

Burke, J. *The Day the Universe Changed*, companion to the PBS Television Series, Little Brown, New York, 1985.

Burke, J. *Connections*, Little Brown, New York, 1978.

Bush, V. "As We May Think," *The Atlantic Monthly*, July 1945.

Butter, A. and Pogue, D. *Piloting Palm*, John Wiley & Sons, Hoboken, NJ, 2002.

Campbell, J. *The Hero with a Thousand Faces,* Pantheon Books, NY, 1949.

Cawsey, A. *The Essence of Artificial Intelligence*, Prentice Hall, Boston, MA, 1998.

Cerf, V. "Information Sharing in the 21st Century," September 2006. www.sefora.org.

Cerf, V. G. and Kahn, R. E. "A Protocol for Packet Network Intercommunication," *IEEE Transactions on Communications* 22(5):1974.

Cervantes, M. "Scientists and Engineers: Crisis, What Crisis?" *Technology and Industry*, January 2004.

Cervone, F. "W3C Delivers Standards for the 'Semantic Web'," Infotoday.com, February 16, 2004. http://newsbreaks.infotoday.com/nbreader.asp?ArticleID=16514.

Childs, P. E. "Chemistry in Action, Chemistry and Chance: Part I." www.ul.ie/~childsp/cinA/issue50/chance.html.

Clarke, R. "A Contingency Approach to the Application Software Generations," Xamax Consultancy Pty Ltd, Canberra, February 1991. http://www.anu.edu.au/people/Roger.Clarke/SOS/SwareGenns.html.

"Claude Shannon, Father of Information Theory, Dies at 84," Bell Labs article, February 26, 2001. www.bell-labs.com/news/2001/february/26/1.html.

CIO Magazine, "Inventing the Enterprise," January, 1, 2000. www.cio.com/archive/010100/metcalfe.html.

Conot, R. E. *Thomas A. Edison: A Streak of Luck*, Da Capo Press, New York, 1979.

Coughlin, K. Claude Shannon obituary entitled "Bell Labs Digital Guru Dead at 84—Pioneer Scientist Led High-Tech Revolution," *The Star-Ledger*, February 27, 2001.

Cringely, R. X. "The Google Box: Taking Over the World Four Ounces at a Time," November 24, 2005. pbs.org/cringely/archive.

Cringely, R. *Accidental Empires: How the Boys of Silicon Valley Make Their Millions, Battle Foreign Competition, and Still Can't Get a Date*, Reprint edition, HarperCollins, New York, 1996.

"Cryptography and Liberty 1998, An International Survey of Encryption Policy," February 1998. http://www.gilc.org/crypto/crypto-survey.html.

The DAML Services Coalition: Anupriya Ankolenkar, Massimo Paolucci, Terry Payne, Katia Sycara, Ora Lassila, Sheila McIlraith, Tran Cao Son, Honglei Zeng, Jerry Hobbs, David Martin, Srini Narayanan, and Drew McDermott, "DAML-S: Web Service Description for the Semantic Web," 2002. http://www.daml.org/services.

Darringer, J. A. et al. "LSS: A System for Production Logic Synthesis," *IBM Journal of Research and Development*, 28(5):537–545, 1984.

"Databases from Scratch I: Introduction." http://brebru.com/databases_from_scratch_1.

Dean, M., Connolly, D., van Harmelen, F., Hendler, J., Horrocks, I., McGuinness, L., Patel-Schneider, P., and Stein, L. "OWL Web Ontology Language 1.0," July 2002. www.w3.org/TR/owl-ref/.

De Bono, E. *Lateral Thinking: Creativity Step by Step*, Harper Paperbacks, New York, 1973.

De Borchgrave, A. "Commentary: Living Forever," United Press International article, Washington, December 29, 2005. http://www.upi.com/HealthBusiness/view.php?StoryID=20051229-090610-4704r.

Descartes, R. *The Philosophical Writings of Descartes (Volume I)*, Cambridge University Press, Cambridge, UK, 1985.

Dertouzous, M. L. "The Future of Computing," *Scientific America*, August 1999.

Dertouzos, M. L. *The Unfinished Revolution: Human-Centered Computers and What They Can Do for Us*, HarperCollins, New York, 2001.

Ding, L., Finin, T., Joshi, A., Peng, Y., Scott Cost, R., Sachs, J., Pan, R., Reddivari, P., and Doshi, V. "Swoogle: A Semantic Web Search and Metadata Engine," Department of Computer Science and Electronic Engineering, University of Maryland, 2004.

"DNA Computing Targets West Nile Virus, Other Deadly Diseases," *Physorg.com*, October 16, 2006. http://www.physorg.com/news80230272.html.

Draper, S. S-Shaped Curve for Uptake of Innovations, 2004.

Duderstadt J. J. "The Future of the University: A Perspective from the Oort Cloud," Emory University Futures Forum, Atlanta, GA, March 8, 2005.

Dyson, G. "Turing's Cathedral," October 24, 2005. www.edge.org.

Edstrom, J. and Eller, M. *Barbarians Led by Bill Gates*, Henry Holt and Company, New York, 1998.

Elon University and the Pew Internet Project, "Imagine the Internet in 2020: A History and Forecast," 2006. www.elon.edu/predictions/.

Englebart, D. "Augmenting Human Intellect: A Conceptual Framework," Stanford Research Institute Summary Report AFOSR-3233, October 1962.

eTForecasts. "Worldwide PC Market," www.etforecosts.com/products/ts_pcww1203.htm.

Fairley, P. "Special Report: R&D '04 By Technology Review Single-Electron Transistors," 2004.

Fensel, D. "The Semantic Web and Its Languages," *IEEE Intelligent Systems* 15(6):67–73, 2000.

Fergusen, A. "The History of Computer Programming Languages." http://www.princeton.edu/~ferguson/adw/programming_languages.shtml.

Friedel, R., Isreal, P., and Finn, B. S. *Biography of an Inventor*, Rutgers University Press, New Brunswick, NJ, 1987.

Frieder, E. and Joyce, K. "Great Ideas: Raymond Kurzweil Receives the World's Largest Award for Innovation," *Spectrum* XIII(3):Fall 2001.

Forrest, D. "A Black Hole," June 2001. http://www.innovationwatch.com/connections.2001.06.00.htm.

Fuller, B. "Quote DB," www.quotedb.com/quotes/107.

Gershenfeld, N. *When Things Start to Think*, Henry Holt and Company, New York, 1998.

Gibbs, W. W. "Trends in Computing," *Scientific American*, 1994.

Gnedenko, B. V. and Kolmogorov, A. N. *Limit Distributions for Sums of Independent Random Variables*. Addison Wesley, Boston, MA, 1954.

Griffin, S. "Internet Pioneers: Douglas Engelbart," 2005. www.ibiblio.org/pioneers/englebart.html.

Gödel, K. "Consistency of the Axiom of Choice and of the Generalized Continuum-Hypothesis with the Axioms of Set Theory," in *Annals of Mathematics Studies*, No. 3, University Press, Princeton, NJ, 1940.

Good, I. J. "Speculations Concerning the First Ultraintelligent Machine," in *Advances in Computers*, Vol. 6, pp. 31–88, Academic Press, San Diego, CA, 1965.

Google Corporate Info, 2006. http://www.google.com/corporate/history.html.

Halstead, M. H. *Elements of Software Science, Operating, and Programming*, Systems Series, Vol. 7, Elsevier, New York, 1977.

Hanson, D. *The New Alchemists: Silicon Valley and the Micro-Electronic Revolution*, Avon, New York, 1982.

Hawkins, J. with Blakeslee, S. *On Intelligence*, Times Books, New York, 2004.

Henry, G. M. "Can We Talk?" *Time Magazine* 127(17):April 28, 1986.

Hiemstra, G. *Turning the Future Into Revenue: What Business and Individuals Need to Know to Shape Their Futures*, John Wiley & Sons, Hoboken, NJ, 2006.

"The History of the MS-DOS Operating Systems, Microsoft, Tim Paterson, and Gary Kildall." http://inventors.about.com/library/weekly/aa033099.htm.

Hoare, C. A. R. "World of Computer Science on Charles Antony Richard Hoare," 2005. en.wikipedia.org/wiki/C._A._R._Hoare.

Hoare, C. A. R. "The Emperor's Old Clothes," The 1980 ACM Turing Award Lecture, *Communications of the ACM* 24(2):February 1981.

Hofstadter, D. R. *Gödel, Escher and Bach: An Eternal Braid*, Basic Books, New York, 1979.

Hook, A. "C# Semantic Search Algorithm," 2003. http://www.headmap.com.

Horrocks, I. and Patel-Schneider, P. "A Proposal for an OWL Rules Language," in *Proceedings of the Thirteenth International World Wide Web Conference (WWW 2004)*, pp. 723–731, ACM Press, New York, 2004.

"Hotwiring Your Search Engine," *Newsweek*, December 19, 2005.

"Imagining the Internet: A History and Forecast," Elon University/Pew Internet Project. http://www.elon.edu/e-web/predictions/150/2010.xhtml.

International Technology Roadmap for Semiconductors (ITRS). www.itrs.com.

"Inventor of the Web Explains Next-Gen 'Semantic Web'," *Marketing VOX Daily*, August 3, 2005.

Jackson, T. *Inside Intel: Andy Grove and the Rise of the World's Most Powerful Chip Company*, The Penguin Group, New York, 1997.

"Jeff Hawkins Q&A," *Technology Review*, July 1999. http://www.technologyreview.com/magazine/jul99/qa.asp.

Juliussen, E. "eTForecasts." www.etforecasts.com/products/ES_pcww1203.htm.

Kahn, H. and Wiener, A. *The Year 2000, A Framework for Speculation on the Next Thirty-Three Years*, MacMillan Publishing Co. New York, 1967.

Kanellos, M., "Intel Scientists Find Wall for Moore's Law," *CNET News.com*, published by ZDNet News, Dec. 1, 2003.

Kay, A. C. "The Early History of Smalltalk," in *History of Programming Languages-II*, ACM Press, New York, 1996.

Kelleher, R. "Google vs. Gates," *Wired* 12(03): March 2004.

Kelly, S. "Intel Looks Beyond the Microchip," BBC News, February 12, 2006.

Kevin, K. "Google vs. Gates," *Wired*, 12(03):March 2004.

Kirsner, S. "The Legend of Bob Metcalfe," *Wired* 6(11):November 1998.

Kleinrock, L. "Information Flow in Large Communication Nets," *RLE Quarterly Progress Report*, July 1961.

Kottke, J. "To Google or Not to Google?" February 26, 2003. kottke.org.

Kurowski, S. "Research Project, 2006." www.scottkurowski.com.

Kurzweil, R. *The Singularity Is Near*, Viking, New York, 2005.

Kurzweil, R. "The Law of Accelerating Returns," published on KurzweilAI.net, March 7, 2001. http://www.kurzweilai.net/meme/frame.html?main=/articles/art0134.html?

Kurzweil, R. Interviews, author of *The Singularity Is Near*, 2005, and with Eliezer Yudkowsky, director of the Singularity Institute for Artificial Intelligence.

Kurzweil, R. "The Lives and Death of Moore's Law," published on KurzweilAI.net September 23, 2003.

Kurzweil, R. "Testimony of Ray Kurzweil on the Societal Implications of Nanotechnology," Committee on Science, U.S. House of Representatives, Hearing, April 9, 2003.

Kurzweil, R. "The Coming Merger of Biological and Non Biological Intelligence," Keynote speech at SC06: the International Conference on High Performance Computing, Networking and Storage, Tampa, FL.

Kurzweil Technologies, Inc. "Biography." www.kurzweiltech.com.

Kurzweil, R. and Meyer, C. "Understanding the Accelerating Rate of Change." www.kurzweiltech.com.

Lacy, M. "An Introduction to Genetic Algorithms in Java." http://www2.sys-con.com/ITSG/virtualcd/Java/archives/0601/lacy/index.html.

Lammers, S. (ed.). *Programmers at Work*, Microsoft Press, Redmond, WA, 1986.

Lampson, B. "Biographic Sketch and Interview of Butler Lampson." http://research.microsoft.com/Lampson/37a-ProgAtWork/37a-ProgAtWork.htm.

Landay, J. A. and Borriello, G. "Design Patterns for Ubiquitous Computing," *Invisible Computing,* August 2003.

Lassila, O. and Swick, R. "Resource Description Framework (RDF) Model and Syntax Specification," W3C Recommendation, World Wide Web Consortium, February 1999. April 11, 2001. www.w3.org/TR/REC-rdf-syntax.

Ledgard, D. "25 Years Since the First Microcomputer—the Altair," June 1999. http://www.colonization.biz/me/altair.htm.

Lee, T. H. "A Vertical Reap in Microchips," *Scientific American.com*, Jan. 2008.

Licklider, J. C. R. and Taylor, R. W. "The Computer as a Communication Device," *Science and Technology*, April 1968.

Lightman, A. and Rojas, W. *Brave New Unwired World*, John Wiley & Sons, Hoboken, NJ, 2002.

Linzmayer, O. W. "30 Pivotal Moments in Apple's History, Highlights—and Lowlights—From the Company's History," 2005.

Lucent Bell Labs, "Claude Shannon, Father of Information Theory, Dies at 84." www.bell-labs.com/news/2001/february/26/1.html.

Lu, S., Dong, M., and Fotouhi, F. "The Semantic Web: Opportunities and Challenges for Next-Generation Web Applications," *Information Research* 7(4):July 2002.

Ma, J., Yang, L. T., Apduhan, B. O., Huang, R., Barolli, L., and Takizawa, M. "Towards a Smart World and Ubiquitous Intelligence: A Walkthrough from Smart Things to Smart Hyperspaces and UbicKids," Japan, February 3, 2005.

Ma, J. et al. "A Walkthrough from Smart Spaces to Smart Hyperspaces Towards a Smart World with Ubiquitous Intelligence," in *11th International Conference on Parallel and Distributed Systems (ICPADS'05)*, pp. 370–376, 2005.

Malik, O. "Free Wi-Fi? Get Ready for GoogleNet," *Business* 2.0, September 2005.

"Marvin Minsky on Common Sense and Computers That Emote," as artificial intelligence research celebrates its 50th birthday, the MIT icon asks What Makes the Minds of Three-Year-Olds Tick?" `www.technologyreview.com/read_article.aspx?id=17164&ch=infotech`.

Mayor, T. "Inventing the Enterprise," *CIO Magazine,* December 15, 1999–January 1, 2000 issue.

Maxfield and Montrose, "1883 AD to 1906 AD The Invention of the Vacuum Tube." `www.maxmon.com/1883ad.htm`.

McCool, R., Fikes, F., and McGuinness, D., "Semantic Web Tools for Enhanced Authoring," Knowledge Systems Laboratory, Computer Science Department, Stanford University, 2003.

McNamee, G. "Amazon Editorial Book Review." `www.amazon.com/Connections-James-Burke/dp/0316116726`.

McIlraith, S., Son, T. C., and Zeng, H. "Semantic Web Service," *IEEE Intelligent Systems* 16(2):46–53, 2001.

Metcalf, R. M. and Boggs, D. R. "Ethernet: Distributed Packet Switching for Local Computer Networks," *Communications of the ACM* 19(7):395–404, July 1976.

Minsky, M., "Steps Toward Artificial Intelligence," *Computers & Thought*, 40–50, 1995.

Minsky, M. `http://web.media.mit.edu/~minsky/minskybiog.html`.

Minsky, M. L. "Matter, Mind and Models," in *Proceedings of the International Federation of Information Processing Congress*, Vol. 1, pp. 45–49, 1965.

Minsky, M. and Papert, S. *Perceptrons*, The MIT Press, Cambridge, MA, 1969.

Minsky, M. "Framework for Representing Knowledge," MIT-AI Laboratory Memo 306, June 1974. Reprinted in *The Psychology of Computer Vision*, P. Winston (ed.), McGraw-Hill, New York, 1975.

"MIT Project Oxygen: Pervasive, Human-Centered Computing." `http://www.oxygen.lcs.mit.edu/`.

MIT, AI Labs. `http://oxygen.lcs.mit.edu/ 2001`.

Moore, G., "Cramming More Components onto Integrated Circuits," *Electronics*, 38(8): April 19, 1965.

Moravac, H. "When Will Computer Hardware Match the Human Brain?" *Journal of Transhumanism* 1: March 1998.

"Nanotechnology," *Encyclopedia Britannica Online*, March 9, 2007, `www.britconnica.com`.

Naudts, G. "An Inference Engine for RDF." `http://www.agfa.com/w3c/2002/02/thesis/An_inference_engine_for_RDF.html`.

Negroponte, N. "MIT's Famed Media Lab." `http://www.technologyreview.com/`.

"The Next Frontier: Google Eyeing a Move Into TV's Territory," *Advertising Age*, 2005. www.macos.utah.edu/Documentation/MacOSXClasses/macosxone/gui.htm.

Noy, N. F. and McGuinness, D. L. "Ontology Development 101: A Guide to Creating Your First Ontology," Stanford University, Stanford, CA, 2000. protege.stanford.edu/publications/ontology_development/ontology101-noy-mcguinness.html.

O'Connor, J. J. and Robertson, E. F. "Kurt Gödel ," *MacTutor History of Mathematics*, October 2003. www-history.mcs.st-andrews.ac.uk/Biographies/Godel.html.

O'Connor, J. J. and Robertson, E. F. "Alan Mathison Turing," *MacTutor History of Mathematics*, October 2003. http://www-history.mcs.st-andrews.ac.uk/Biographies/Turing.html.

Ogbuji, U. "The Languages of the Semantic Web," *New Architect*, June 2002.

Olson, S. "Interview of Ray Kurzweil," Center for Nanotechnology Responsibility, December 2005. http://www.crnano.org/interview.kurzweil.htm.

O'Neill, D. "Global Software Competitiveness Studies," Center for National Software, 2004.

Orwell, G. *1984*, 1st World Library, July 2005.

"Out of the Ether," *The Economist*, September 2003. www.economist.com/science/tq/displayStory.cfm?story_id=2019967.

Page, T. *Dancing in the Dark*, Smithsonian, March 2006.

Penn State Research Institute, "Anatomy of Discovery," *Physorg.com* September 29, 2006. www.physorg.com/news78754375.html.

Peterson, H. *A Treasury of the World's Great Speeches*. Spencer Press, Chicago, 1954.

Prather, M. "Ga-Ga for Google," *Entrepreneur Magazine*, April 2002.

PBS, "Shockley, Brattain and Bardeen." www.pbs.org/transistor/album1/addlbios/egos.html.

Pink, D. H. *A Whole New Mind,* Riverhead Books, New York, 2005.

Playboy Interview, "Google Guys," *Playboy*, September 2004.

Popular Electronics, January 1975. http://www.computermuseum.20m.com/popelectronics.htm.

"Privacy and Ubiquitous Network Societies," Background Paper, International Telecommunication Union (ITU) Document: UNS/05, ITU Workshop on Ubiquitous Network Societies, April 2005.

Quittner, J. and Slatalla, M. *Speeding the Net: The Inside Story of Netscape and How It Challenged Microsoft*, Atlantic Monthly Press, New York, 1998.

Redding, C. "Wireless Networking Analysis and Forecasting," Institute for Telecommunication Sciences (ITS), 2005.

Reimer, J. "A History of the GUI," May 5, 2005. arstechnica.com/articles/paedia/gui.ars/1.

Robat, C. (ed.). *Introduction to Software History*. www.thocp.net/software/software_reference/introduction_to_software_history2.htm#ai.

Roberts, R. M. *Serendipity*, John Wiley & Sons, Hoboken, NJ, 1989.

Rogers, E. M. "*Diffusion of Innovations*," 4th ed., The Free Press, New York, 1995, Chap. 1, p. 11, Fig. 1-1.

Roush, W. “Marvin Minsky on Common Sense and Computers That Emote,” July 13, 2006. http://www.techreview.com/Infotech/17164/page1/.

Russell, B. *Introduction to Mathematical Philosophy*, Dover Publications, Mineola, NY, 1993.

Sabbatini, R. M. E. “The Mind, Artificial Intelligence and Emotions,” Universidade Estadual de Campinas, 1998. http://www.cerebromente.org.br/n07/opiniao/minsky/minsky_i.htm.

Scacchi, W. “Understanding Software Productivity,” UCLA, 1994.

Schaeffer, J. “A Gamut of Games,” *AI Magazine*, September 22, 2001.

“Search Us, Says Google,” *Technology Review*, January 11, 2002.

Searle, J. R. “Minds, Brains, and Programs,” *The Behavioral and Brain Sciences, Volume 3*, Cambridge University Press, Cambridge, UK, 1980.

Second International Symposium on Ubiquitous Intelligence and Smart Worlds (UISW2005), Nagasaki, Japan, December 6–7, 2005. www.ubiquitous-intelligence.org/conf/uisw2005.

Segaller, S. *Nerds 2.0.1: A Brief History of the Internet*, TV Books, New York, 1998.

Shankland, S. “IBM: Linux Investment Nearly Recouped,” *CNET News.com*, January 29, 2002.

Shannon, C. E. “A Mathematical Theory of Communication,” *The Bell System Technical Journal*, 27:379–423, 623–656, July, October, 1948.

Shannon, C. E. *A Symbolic Analysis of Relay and Switching Circuits*, Master’s thesis, Massachusetts Institute of Technology, 1940.

Simonyi, C. *Meta-programming: a Software Production Method*, Ph.D. thesis, Stanford University, December 1976. http://www.parc.xerox.com/publications/bw-ps-gz/csl76-7.ps.gz (December 2001).

Smart, J. “Critiques of Prediction,” Los Angeles, CA, 2006. www.accelerationwatch.com/critiques.html.

Smith, D. and Alexander, R. *Fumbling the Future,* Holt and Co., New York, 1988.

“Steven P. Jobs and Stephen Wozniak, Apple Computer, The Personal Computer Is Born,” 1985. http://www.thetech.org/nmot/detail.cfm?ID=17&STORY=3&.

Softky, M. “Building the Internet,” *The Almanac*, October 11, 2000. http://www.almanacnews.com/morgue/2000/2000_10_11.taylor.html.

Softky, M. Almanac Staff, October 2001. http://www.almanacnews.com/morgue/2000/2000_10_11.taylor.html.

Stork, D. G. “Artificial Intelligence in the World Wide Web,” published on KurzweilAI.net, March 7, 2001. http://www.kurzweilai.net/meme/frame.html?main=/articles/art0137.html.

Sutherland, J. “The Ideas Interview: Ray Kurzweil,” *The Guardian*, November 21, 2005.

Taylor, C. “Imagining the Google Future,” February 1, 2006. blog.outer-court.com/archive/2006-02-01-n55.html.

Taylor, P. “Alan Mathison Turing (1912–1954),” Australian Mathematics Trust, May 14, 2002. http://www.amt.canberra.edu.au/turingb.html.

Teevan, J., Alvarado, C., Ackerman, M., and Karger, D. "The Perfect Search Engine Is Not Enough: A Study of Orienteering Behavior in Directed Search" CSAIL, Massachusetts Institute of Technology, Cambridge, MA, 2005.

"The Google Browser," August 24, 2004. kottke.org.

"The Open Mind Initiative." http://www.openmind.org/.

"The Ethernet Effect: Collaboration, Interoperability and Adoption of New Technologies," A University of New Hampshire InterOperability Laboratory White Paper in Collaboration with Dell'Oro Group, April 2006. http://www.iol.unh.edu/services/testing/fe/training/The_Ethernet_Effect_WhitePaper.pdf.

Toffler, A. *Future Shock,* Random House, New York, 1970.

Torvalds, L. *Linux: A Portable Operating System,* Master's thesis, University of Helsinki, 1996.

http://tripletsandus.com/80s/commercials.htm.

Tripathi, A. K. "Reflections on Challenges to the Goal of Invisible Computing," *Ubiquity* 17:May 17–24, 2005.

Turing, A. M. "On Computable Numbers, With an Application to the Entscheidungsproblem," *Proceedings of the London Mathematics Society* 2(42):1936.

Turing, A. M. "Computing Machinery and Intelligence," *Mind* 49: 433–460, 1950.

Tuomi, I. "The Life and Death of Moore's Law," *First Monday*, 7(11): November, 2002. www.firstmonday.org/issues/issue7_11/tuomi/.

"UCLA Chemists, Hewlett-Packard Labs Colleagues Report Significant Advances Toward Chemical Computers." www.sciencedaily.com/releases/1999/07/990719082047.htm.

"Ubiquitous Computing," Xerox Palo Alto Research Center—Sandbox Server. http://sandbox.xerox.com/ubicomp/.

Updegrove, A. "A New Vision from the Inventor of the World Wide Web: An Interview with Tim Berners-Lee."

U.S. Congress, "National Defense Education Act," *Public Law 85-864*, 1958.

U.S. Department of Defense, DoD directive 5105.15 establishing the Advanced Research Projects Agency (ARPA), signed on February 7, 1958.

"Vacuum Tube Valley," August 8, 2001. www.vacuumtube.com/toppage11.htm.

van der Werff, T. J. "Red Herring's Top Ten Trends 2003." www.globalfuture.com/redh-trends2003.htm.

Vise, D. A. and Malseed, M. *The Google Story*, Delacorte Press, New York, 2005.

Von Neumann, J. *The Computer and the Brain*, Yale University Press, New Haven, CT, 1948.

Wallace, J. and Erickson, J. *Hard Drive: Bill Gates and the Making of the Microsoft Empire*, HarperCollins, New York, 1993.

Wallace, J. *Overdrive: Bill Gates and the Race to Control Cyberspace.* John Wiley & Sons, Hoboken, NJ, 1997.

Wang, P. "A Search Engine Based on the Semantic Web," M.Sc. in Machine Learning and Data Mining Project, University of Bristol, May 2003.

Warner, B. "Microchip." technology.timesonline.co.uk/article/0,,20749-1555044,00.html.

Warner, W. "Great Moments in Microprocessor History. The History of the Micro from the Vacuum Tube to Today's Dual-Core Multithreaded Madness," December 22, 2004. www-128.ibm.com/developerworks/library/pa-microhist html?ca=dgr-lnxw01MicroHistory.

Weiser, M. "The World Is Not a Desktop," November 7, 1993. www.ubiq.com/hypertext/weiser/ACMInteractions2.html.

Weiser, M. "The Computer of the 21st Century," *Scientific American*, September 1991. http://www.ubiq.com/hypertext/weiser/SciAmDraft3.html.

Weiser, M. and Brown, J. S. "Designing Calm Technology," Xerox PARC, December 21, 1995. http://sandbox.xerox.com/hypertext/weiser/calmtech/calmtech.htm.

Weisman, R. "MIT Is Readying New Technologies that Put Humans in the Center of Computing," MIT Globe Staff, June 21, 2004.

Weisstein, E. "Eric Weisstein's World of Biography." scienceworld.wolfram.com/biography/Shannon.html.

Weisstein, E. W. "Goldbach Conjecture," from *MathWorld*—A Wolfram Web Resource, updated 2007. http://mathworld.wolfram.com/GoldbachConjecture.html.

Wikipedia, the free encyclopedia, 2006. en.wikipedia.org/wiki/.

Wilde, O. "Quote DB." http://www.quotedb.com/quotes/111.

Wolfram, S. "A New Kind of Science," 2002.

"World's First Minicomputer Kit to Rival Commercial Models," *Popular Electronics*, January 1975.

Wu, J. Xerox PARC December 5, 2002. www.quad4x.net/cswebpage/parc.html. www.anu.edu.au/people/Roger.Clarke/SOS/SwareGenns.html.

"Xerox Names Computing Pioneer as Chief Technologist for Palo Alto Research Center," Xerox press release, August 14, 1996. http://www.ubiq.com/weiser/weiserannc.htm.

Yen, D. "Perspective: End of Moore's Law? Wrong Question," February 18, 2004. news.com.com/2010-1006-5160336.html.

Supplemental Reference Web Sites

www.answers.com/topic/alexander-grarham-bell?cat=technology.

www.bell-labs.com/news/2001/february/26/1.html.

www.colonization.biz/me/altair.htm.

www.cio.com/archive/010100/metcalfe.html.

www.hp.com/hpinfo/newsroom/feature_stories/2002/alankaybio.html.

www.ibiblio.org/pioneers/englebart.html.

www.kurzweilai.net/meme/frame.html?main=/articles/art0353.html?

www.kurzweilai.net/bios/frame.html.

research.microsoft.com/lampson/37a-ProgAtWork/37a-ProgAtWork.htm.

www.quotationspage.com/quote/306.html.

www.quotedb.com/quotes/107.

www.sec.gov/Archives/edgar/data/1288776/000119312504139655/ds1a.htm, Google's S-1 registration statement.

www.smalltalk.org/alankay.html.

www.sysprog.net/quotdesi.html.

www.marketingprofs.com/login/join.asp?adref=rdblk&source=/5/updegrove1.asp.

secondlanguagewriting.com/explorations/Archives/2006/Oct.html, an interview in *Technology Review* via elearnspace), stated: Sunday, October 1, 2006 3:03 PM.

www.technologyreview.com/InfoTech/12219/page3/, and many other sites; interview by *Technology Review* entitled "Search Us, Says Google."

www.thestreet.com/_yahoo/markets/marketfeatures/10157519_6.html, Text of Google's Letter to Prospective Shareholders.

thinkexist.com/quotation/the-ultimate-search-engine-would-understand/695067.html, and also included in numerous other web sites and printed publications.

www.wired.com/wired/archive/12.03/google.html?pg=13, and other sites; quoted in *Wired* article "Google vs. Gates" by K. Kelleher.

www.wired.com/wired/archive//6.11/metcalfe_pr.html.

www.w3.org/2005/Talks/0621-dsr-ubiweb/#(1).

Glossary

agent A piece of software that runs without direct human control or constant supervision to accomplish a goal provided by the user. Agents typically collect, filter, and process information found on the Web, sometimes in collaboration with other agents.

analog Refers to an electronic device that uses a system of unlimited variables to measure or represent flow of data. Radios use variable sound waves to carry data from transmitter to receiver.

applet A small software application or utility that is built to perform one task over the Web.

backbone The largest communications lines on the Internet that connect cities and major telecommunication centers.

bandwidth The carrying capacity or size of a communications channel; usually expressed in hertz (cycles per second) for analog circuits and in bits per second (bps) for digital circuits.

browser A Web client that allows a human to read information on the Web. Microsoft Internet Explorer and Netscape Navigator are two leading browsers.

CERN (Conseil Européen pour la Recherche Nucléaire) European Particle Physics Laboratory of the European Orgaization for Nuclear Research. The European Particle Physics Laboratory, located on the French–Swiss border near Geneva, Switzerland.

Connections: Patterns of Discovery By H. Peter Alesso and Craig F. Smith
Copyright © 2008 John Wiley & Sons, Inc.

client Any program that uses the service of another program. On the Web, a Web client is a program, such as a browser, editor, or search robot, that reads or writes information on the Web.

cwm (closed world machine) A bit of code for playing with this stuff, as grep is for regular expressions. Sucks in RDF in XML or N3, processes rules, and spits it out again.

Cyc A knowledge-representation project that expresses real-world facts in a machine-readable fashion.

DAML (DARPA Agent Markup Language) The DAML language is being developed as an extension to XML and the Resource Description Framework (RDF). The latest release of the language (DAML+OIL) provides a rich set of constructs with which to create ontologies and to markup information so that it is machine readable and understandable. `http://www.daml.org/`.

decentralized network A computer network distributed across many peers rather than centralized around a server.

digital An electronic device that uses a predetermined numbering system to measure and represent the flow of data. Modern computers use digital 0's and 1's as binary representations of data.

distributed artificial intelligence (DAI) DAI is concerned with coordinated intelligent behavior: intelligent agents coordinating their knowledge, skills, and plans to act or solve problems, working toward a single goal, or toward separate, individual goals that interact.

expert system A computer program that has a deep understanding of a topic and can simulate a human expert, asking and answering questions and making decisions.

eXtensible Markup Language (XML) Separates content from format, thus letting the browser decide how and where content gets displayed. XML is not a language, but a system for defining other languages so that they understand their vocabulary.

HTML (Hypertext Markup Language) A computer language for representing the contents of a page of hypertext; the language that most Web pages are written in.

HyperLink See *Link hypertext—nonsequential writing*, Ted Nelson's term for a medium that includes links. Today it includes other media apart from text and is sometimes called hypermedia.

HyperText Transfer Protocol (HTTP) This is the protocol by which Web clients (browsers) and Web servers communicate. It is stateless, meaning that it does not maintain a conversation between a given client and server, but it can be manipulated using scripting to appear as if state is being maintained. Do not confuse HTML (markup language for our browser-based front ends) with HTTP (protocol used by clients and servers to send and receive messages over the Web).

hub A point where communications lines are brought together to exchange data.

Hyperlink Elements such as text, graphics, and other objects embedded in a Web page's HTML code that establishes connections to related Web pages or elements.

hypernavigation Occurs when a rendering plug-in directs the client to display a URL at a specified time in a stream. When the plug-in issues a hypernavigation request, the default Web browser opens.

Internet A global network of networks through which computers communicate by sending information in packets. Each network consists of computers connected by cables or wireless links.

Intranet A part of the Internet or part of the Web used internally within a company or organization.

IP (Internet Protocol) The protocol that governs how computers send packets across the Internet; designed by Vint Cerf and Bob Khan.

International Standards Organization (ISO) A nontreaty standards organization active in development of open systems interconnections.

Internet Service Provider (ISP) A company that lets users dial into its computers that are connected to the Internet.

Internet Protocol address (IP address) The numeric address used to locate computers on a TCP/IP network. The numbers include four groups each separated by a period.

Java A programming language developed (originally as *Oak*) by James Gosling of Sun Microsystems. Designed for portability and usability embedded in small devices, Java took off as a language for small applications (*applets*) that ran within a Web browser.

kbps kilobytes per second.

knowledge base An informal term for a collection of information that includes an ontology as one component. Besides an ontology, a knowledge base may contain information specified in a declarative language such as logic or expert-system rules, but it may also include unstructured or unformalized information expressed in natural language or procedural code.

knowledge discovery The process of complex extraction of implicit, previously unknown, and potentially useful knowledge from large datasets. Coined in 1989 by artificial intelligence and machine learning researchers.

knowledge management The process of creating, capturing, and organizing knowledge objects. A knowledge object might be a research report, a budget for the development of a new product, or a video presentation. Knowledge management programs seek to capture objects in a repository that is searchable and accessible in electronic form.

kps kilobytes per second, a measure of the data rate. See **kbps**.

learning The process of automatically finding relationships between inputs and outputs given examples of that relationship.

link A link (or hyperlink) is a relationship between two resources. HTML links usually connect HTML documents together in this fashion (called a *hyperlink*), but links can link to any type of resource (documents, pictures, sound, and video files) capable of residing at a Web address.

markup language Ued to structure a document's character data into logical components, and "name" them in a manner that is useful. These labels (element names) provide either formatting information about how the character data should be visually presented (e.g., for a word processor or a Web browser) or they can provide semantic (meaningful) information about what kind of data the component represents. Markup languages provide a simple format for exchanging text-based character data that can be understood by both humans and machines.

metadata Data about data on the Web, including but not limited to authorship, classification, endorsement, policy, distribution terms, IPR, and so on. A significant use for the Semantic Web.

Meta-markup language A language used to define markup languages. SGML and XML are both meta-markup languages. HTML is a markup language that was defined using the SGML meta-markup language.

natural language processing (NLP) Using software to "understand" the meaning contained within texts. Everyday speech is broken down into patterns. Typically, these systems employ syntactic analysis to infer the semantic meaning embedded in documents. NLP identifies patterns in sample texts and makes predictions about unseen texts. Also called computational linguistics.

object A unique instance of a data structure defined according to the template provided by its class. Each object has its own values for the variables belonging to its class and can respond to the methods defined by its class.

ontologies Collections of statements written in a language such as RDF that define relationships between concepts and specific logic rules. Semantic data on the Web will be understandable by following the links to specific ontologies.

OWL Web Ontology Language for markup ontology for the Internet.

OWL-S Web Ontology Language for Services.

RDF (Resource Description Framework) A framework for constructing logical languages that can work together in the Semantic Web. A way of using XML for data rather than just documents.

Semantic Web Communication protocols and standards that would include descriptions of the item on the Web such as people, documents, events, products, and organizations, as well as relationships between documents and relationships between people.

Semantic Web Services Web services developed using semantic markup language ontologies.

server A program that provides a service (typically information) to another program, called the client. A Web server holds Web pages and allows client programs to read and write them.

SGML (Standard Generalized Markup Language) An international standard in markup languages, a basis for HTML and a precursor to XML.

spider (crawler) A spider is a program that browses web sites, extracting information for search engine database. Spiders can be summoned to a site through search engine registration or they will eventually find your site by following links from other sites (assuming you have links from other sites).

stemming The removal of suffixes and sometimes prefixes from words to arrive at a core that can represent any of a set of related words.

syntactic The part of language concerned with syntax, sentence structure. For example, the phrases "my mother's brother" and "my uncle" express the same relationship, but the way in which the information is expressed differs.

Structured Query Language (SQL) An ISO and ANSI standard language for database access. SQL is sometimes implemented as an interactive, command line application and sometimes is used within database applications. Typical commands include select, insert, and update.

taxonomy This term traditionally refers to the study of the general principles of classification. It is widely used to describe computer-based systems that use hierarchies of topics to help users sift through information. Many companies have developed their own taxonomies, although there are also an increasing number of industry standard offerings. Additionally, a number of suppliers, including Applied Semantics, Autonomy, Verity, and Semio, provide taxonomy-building software.

Transmission Control Protocol/Internet Protocol (TCP/IP) Two protocols used together to govern communication between Internet computers. HTTP (HyperText Transfer Protocol) uses TCP as the protocol for reliable document transfer. If packets are delayed or damaged, TCP will effectively stop traffic until either the original packets or backup packets arrive. Among the tools that enabled the development of the Internet and the subsequent explosive growth of the World Wide Web is TCP/IP, a suite of network communications protocols used to connect hosts on the Internet. TCP/IP is comprised of several protocols, the two main ones being TCP and IP. TCP/IP has become the de facto standard for transmitting data over networks. Even network operating systems that have their own protocols, such as Netware, also support TCP/IP.

Universal Resource Identifier (URI) A URI defines an entity. URLs are a type of URI.

Universal Resource Locator (URL) The familiar codes (such as `http://www.sciam.com`) that are used as hyperlinks to Web sites.

W3C (World Wide Web Consortium) A neutral meeting of those to whom the Web is important, with the mission of leading the Web to its full potential. The World Wide Web Consortium (W3C) is an organization that was founded in October 1994 as a forum for information exchange, commerce, communication, and collective education. The W3C is comprised of individuals and organizations located all over the world and involved in many different fields. Members participate in a vendor-neutral forum for the creation of Web standards. W3C develops interoperable technologies (specifications, guidelines, software, and tools) intended to enable further development of the World Wide Web and lead it to its full potential.

Web services Web-accessible programs and devices.

Web server A program that, using the client/server model and the World Wide Web's HyperText Transfer Protocol (HTTP), serves the files that form Web pages to Web users (whose computers contain HTTP clients that forward their requests).

XML XML stands for eXtensible Markup Language. The key feature of XML in comparison with HTML is that it provides the ability to define tags and attributes, not allowed under HTML. XML is a subset of the Standard Generalized Markup Language (SGML) designed for use on the Internet. It supports all the features of SGML and valid XML documents are therefore valid SGML documents.

XSDL XML Schema Description Language is the W3C recommendation that goes beyond DTD with the addition of XML datatypes, namespace support, and inheritance mechanisms.

XML Schema A formal definition of a "class" or "type" of documents that is expressed using XML syntax instead of SGML DTD syntax.

Index

Connections: Patterns of Discovery By H. Peter Alesso and Craig F. Smith
Copyright © 2008 John Wiley & Sons, Inc.